SWIFTS IN A TOWER

SWIFTS IN A TOWER

Dr David Lack
With a new chapter by Dr Andrew Lack

UNICORN

Published in 2018 by
Unicorn, an imprint of Unicorn Publishing Group LLP
5 Newburgh Street
London
W1F 7RUG
www.unicornpublishing.org

First published 1956 by Methuen & Co Ltd
Reprint 1973 published by Chapman and Hall Ltd
11 New Fetter Lane, London EC4P

ISBN 978-1-911604-36-5

10 9 8 7 6 5 4 3

Typeset by Vivian@Bookscribe

Printed and bound in India by Imprint Press

CONTENTS

PLATES

Plate 1. Adult swift incubating.
Plate 2. Adult near the entrance hole to its nest, showing the eye set deep in the head for protection in fast flight.
Plate 3. Adult flying showing the pale throat patch.
Plate 4. Adult rapidly changing direction in flight showing spread wings and tail raised and fully splayed.
Plate 5. Oxford University Museum. The west and south sides of the tower are visible, each with ten cowls. There are two swift boxes under each cowl. Note that the carvings around the windows were never completed.
Plate 6. A screaming party.
Plate 7. Adult bringing a feather as nest material.
Plate 8. Adult with throat distended as full of a bolus of small insects for the newly hatched young.
Plate 9. Adult feeding newly-hatched young.
Plate 10. Young about seven days old showing the feathers starting to erupt and unopened eyes.
Plate 11. Young about fifteen days old with eyes starting to open and feathers growing.
Plate 12. Two young about fifteen days old with feathers growing and a discarded egg beside the nest.
Plate 13. Well-grown young showing extensive white around the beak. Adult on right.
Plate 14. Well-grown young showing white edges of head and wing feathers.
Plate 15. Adult approaching nest. Photograph taken from the nest entrance.
Plate 16. Adult about to enter nest with wings and tail spread to brake and feet forward.
Plate 17. Young swift showing very short beak but large gape.
Plate 18. Flat fly, *Crataerina pallida* on young swift.
Plate 19. Roy Overall, long-term keeper of the swifts, by the video link in the main court of the University Museum.

FIGURES

BRIEF ENCOUNTER

Three precious months
Is all that they could stay,
May, flaming June
And hot July.
Now swifts have left
To our dismay.

Like anchors
Grappling clouds in seas of sky,
They weighed
To let birds slip away.
We had no way
To say Goodbye.

Peter Brown, from *Swifts Round a Tower* (2014)

FOREWORD

For me the proverb 'One swallow does not make a summer' has never carried much weight. Two swallows might be harbingers of spring, but summer only truly comes when the parties of swifts scream overhead. They may lack the melodiousness of the blackbird singing in the Museum forecourt, but those screams really do tell us that summer has arrived.

The birds do not nest until they have had a spell of really warm weather to get into breeding condition and they leave us as soon as their chicks have flown; prompting David Lack to say that their autumn departure depended on the time of spring.

The end of their breeding season happens suddenly; after some five weeks in the nest the chicks, having been staring out of the nest-hole for a few days, suddenly take the plunge - and they're off. The parent, returning with a bolus of food, finds that there is no one there to feed it to; it is unlikely that they will ever meet again. At least some chicks leave the country almost immediately; one was killed in Madrid just three days after it had left its nest in the Tower. The parents may remain for a few days; the parents of this chick were still in Oxford when it died.

Most young birds have the benefit of practice flights or short sallies into the air; not so the swift, once launched it's in the air day and night for nine months, perhaps much more. Many young birds have seen a little of what they need to feed on; not so the swift, it has never seen a flying insect, just swallowed a crumpled congealed mass of little objects.

It is likely that the numbers of these little objects is the key to the swift's current decline in numbers. There is concern that they may be short of nesting sites (they certainly aren't in the Tower). However, looking more broadly, the other birds that, like them, feed on flying insects – the Swallow, the House Martin and the Sand Martin are all in serious decline. These species have very different nesting requirements, but all rely on flying insects. Ask any senior citizen and they will tell you how they used to have to scrape the insects off the windscreens and bonnets of their cars. Insects in general seem in serious decline, the general view being that the cause lies in the intensification of agriculture and the use of agro-chemicals that has accompanied it.

In some ways we know little more about their daily lives than we did when David Lack wrote his classic book. The new final chapter describes many studies carried out since then; we know in much more detail the migratory routes they take, the heights at which they fly and many other details of their lives. We know much more about what they do, but how they do it is as much of a mystery as it was when *Swifts in a Tower* was written.

Swifts lack a melodious song, and they are dull and black lacking any strikingly coloured plumage; yet to anyone who knows them they are amongst the most remarkable of birds. They were born to fly.

Professor Christopher Perrins

NOTE ON THIS EDITION

This is a full reprint of David Lack's *Swifts in a Tower* published in 1956. I discuss all the updating in the appended chapter that demonstrates how much has been discovered in the intervening 62 years. Nothing has been altered from the original text except for references to the Plates. For this edition we have used new photographs. We felt that photography has improved so much since 1956 that the originals, though excellent and 'unique' at the time, have been superseded. We have kept all the original line-drawn Figures except the maps of Figs. 13 and 18 that we have re-drawn (see my Acknowledgements) with the more accurate information we now have. I have left the original Figs. 20 and 21 as they were, but replacement maps using current information are in the new chapter as Figs. 26 and 27.

This updated edition has been made possible by National Lottery players and funded by the Heritage Lottery Fund, as part of the Oxford Swift City project. This two-year project was launched in May 2017, led by the RSPB, seeking to improve the outlook for swifts in Oxford by raising local awareness of the swifts' plight and the actions people can take to help them.

Andrew Lack

Chapter 1

INTRODUCTION

The swift is one of the most remarkable, yet one of the least known, of all British birds. In flight it is pre-eminent, indeed it spends more of its life in the air than any other bird. But though it breeds freely in our villages and towns, it places its nest in half-darkness under a roof or in a hole in a high wall, so that few have even seen it, let alone studied the breeding birds. Although, too, its shrill scream is a characteristic sound of summer evenings, the swift is scarcely ever mentioned in folk-lore or poetry, and then it has often been confused with the martin; and country-folk, when they know it, have called it the Devil's bird from its black plumage. Even the naturalists have rarely watched it, and when the present study was begun in 1946, the fullest account of its habits was still that written by Gilbert White in the year 1774.

This book is intended for the general reader who wishes to know more of this strange guest under our roofs, and of the problems raised by so aerial a life, hence I have tried to write simply and without technical terms. It is based on ten years' observations, and deals first with the home life of the swift and then with its life in the air. In a more didactic age it might have been subtitled 'The Perils of Specialisation', for the swift's dependence on the air brings many benefits but also a great danger. Its magnificent flight enables it to take all its food and nesting material from the air, to drink and to bathe without alighting, to mate in the air and to spend the night on the wing, while except when weakened or surprised it can out-fly birds of prey. But the air-borne insects upon which it relies for food become extremely scarce in bad weather, and the danger of starvation can be overcome only by unusual behaviour and other adaptations.

A swift is easily distinguished from a swallow or martin as it is much larger, with recurved wings and stronger flight, and it has black while they have white underparts. Indeed, the swift is black all over save for a pale chin, the two sexes being alike in colour and size. Its shrill scream is also characteristic, swallow and martin having a more homely twitter. Swifts reach us later in the spring than swallows, arriving at the beginning of May and leaving in early August, so that they are 'British birds' for only about a quarter of their life. They spend the winter in southern Africa.

In most bird-books the swift is shown perched or crouched on a rock, but in fact it spends almost its whole life either in flight through the wide spaces of the air or cramped in a small dark hole. Except on rare occasions at night, the swift, unlike the swallow and martin, does not settle in the open. Hence the traditional 'bird-book swift' depicted in Fig. 1 is very misleading; further, its feet have been set in the middle of its body instead of in front, and it has craned its head towards the reader, whereas the swift has a short neck and can turn it only a little (Plate 1, see also Plates 15, 16). Most of the pictures of swifts in flight are equally unnatural, their bodies being too heavy, their necks too long and their wings too thin, with too small an area of primary feathers. Many of them would be unable to fly, and at best they are shown with proportions that fit a swallow rather than a swift. The interested reader can compare the true appearance in flight, with the illustrations in standard works on birds shown in Plate 3. These errors reflect the extent to which swifts differ from other birds in their extreme adaptation for moving through the air at high speed.

The bird which forms the subject of this book is really the common swift (*Apus apus*), but for simplicity it will be referred to as the swift, since no other kind breeds in Britain or, for that matter, in northern Europe. Further south in Europe occur two related species, the alpine and pallid swifts, while in North America there is the chimney swift, which is more distantly related and has rather different habits, also it is half the size of our bird. Two other species breed in western North America and two more in northern Asia. All of these, like the European forms, stay in the north only for the summer and migrate south in the autumn to spend the rest of the year in or near the tropics.

The few swifts in cool regions are offset by the many in warmer lands. There are some seventy species in the world, and the tropics of Asia, Africa and America each support about twenty. As many as nine different forms have been seen in one flock in Kenya, eleven have been recorded in one place in Venezuela, while one country, the Belgian Congo, probably has fifteen breeding species. Nearly all these swifts look very alike, being streamlined, slender-winged, high-speed fliers, with short legs, small bills and large mouths. They see each other, as we see them, either in the air or in dark nesting holes, hence colour could have little value in their lives, and nearly all of them are black above and below, usually sooty, though some are glossy and some show patches of white. Even in size they do not greatly vary, the largest having a wing-span of between two and three times the smallest. As a result, they are hard to distinguish from each other in the field, and little is known about the habits of most of them. Pictures of the main types are given in Chapter 5, with an appendix summarising their characteristics and distribution. It is helpful to appreciate that our own swift is one of a large family, and if the reader should feel that in some respects it manages its life badly, he should remember that it belongs to a tropical group, and should, rather, admire the ways in which it has become fitted for our cool, wet climate.

The observations recorded here were made jointly with my wife, but family duties did not allow us to write the book together. Where 'we' is used, it refers to us both, and only a few observations on foreign swifts were made by me alone. Several other members of the Edward Grey Institute at Oxford helped with the observations at times, including I. Werth, P. H. T. Hartley, M. T. Myres, D. F. Owen and P. W. Davies. I am also grateful to H. N. Southern for his unique photographs of swifts in the nesting boxes in the tower, to C. Eric Palmar for three other photographs, to Commander A. M. Hughes for his critical skill in the drawings of foreign swifts, to Mrs B. K. Sladen for another drawing and to Mrs R. Heelas for the maps. J. Barlee read the chapter on flight, and J. Buxton, J. Monk and R. E. Moreau the whole book in manuscript, and gave valuable criticisms which resulted in much rewriting. R. E. Moreau, it should be added, was the first to make an intensive study of swifts at the nest, with the result that, for a time, three African species were better known than our British bird. My biggest debt is to E. Weitnauer, schoolmaster of Oltingen in Switzerland, for the reason given in the next chapter. In addition to these personal friends, there are the many naturalists past and present, who by their published observations on swift and other birds have contributed indirectly to the book. To lighten the text, references to the work of others are relegated to a special appendix, and those who wish to read further, or to check the source of observations made by others, should refer to the appropriate page number in this list.

Gilbert White both studied swifts and wrote verses about them. 'Amusive birds', he called them, as they dashed screaming round Selborne church. I hope, similarly, that the book will convey to the reader both the scientific interest of the bird and the delight to be obtained from just watching it. It calls forth, perhaps, rather little affection, but unbounded admiration. Dr N. Tinbergen has persuasively argued that the study of birds, like bird-photography, is enjoyable because it is an expression of man's hunting instinct. There is the same pleasure in natural surroundings, the same need to overcome difficulties through skill and the same triumph in the trophy. He was doubtless anxious to offset the tendency of an earlier generation to ignore the instinctive basis of human activities, but it may also be dangerous to ignore the higher part. Those who built the tower where our own swifts live held a grander view of science, and if they had not done so, the tower would not have been built and this book would not have been written.

Fig 1: The 'bird-book' swift

Chapter 2

THE TOWER

The story begins on 20 June 1855, when a group of bearded and reverend scientists, a combination now, alas, unknown, joined to sing the Benedicite in the open air on the edge of Oxford, as the foundation stone was laid for a remarkable new building, Oxford's University Museum of Science (see Plate 5). And as they sang 'O, all ye fowls of the air, bless ye the Lord' there came, it may be supposed, an answering scream from the circling swifts, for in the top of the new building these birds would find their home in the years to come.

The museum was built only after fierce argument. The idea that a university should have a special museum for science was hotly disputed, and when the principle was grudgingly accepted, funds were lacking, until it was revealed that £60,000 lay in the University Chest from the Clarendon Press, profits from the sale of bibles. To use money so obtained for the promotion of science was to many sacrilegious, and the proposal would have been rejected by Convocation but for the strong support of Dr Pusey and his fellow members of the High Church party.

There was further argument as to the architecture which would be worthy of so novel an object. In an open competition, in which each design was submitted anonymously under a motto, that was eventually chosen which carried 'Nisi Dominus aedificaverit domum' (Except the Lord build the house, their labour is but lost that build it). Gradually the building rose, glowing in its variegated brick and stone, and reflecting after its fashion the old glories of French château and Gothic cathedral, and the new glories of plate glass and steel girder. Inside were placed the fossils, the stuffed animals, the crystals and the mounted insects, together with the last remains of the last dodo, so unhappily burnt by order of a former Vice Chancellor. Surmounting all, a grand but useless ornament, was a tower sixty feet higher than the rest and empty within. But it is no longer useless, for here the swifts have made their homes – the living above the dead (Plate 5).

The spirit in which the new building went forward is shown by the prayer composed for the foundation ceremony by the Professor of Medicine, Sir Richard Acland. 'Grant that the building now to be erected on this spot may foster the progress of those sciences which reveal to us the wonders of Thy creative powers. And do Thou, by Thy heavenly grace, cause the knowledge thus imparted to fill us with the apprehension of Thy greatness, Thy wisdom and Thy love.' In those days, scientists knew that God was glorified in His works. There was no forewarning shadow of the great debate on Darwinism, to be held in the scarcely completed building only five years later, at the meeting of the British Association for the Advancement of Science in 1860, when the Bishop of Oxford with misplaced humour challenged an agnostic about his ancestry and T. H. Huxley replied, it is said, that he would rather be descended from an ape than from a divine who used authority to stifle truth. Science became divided from religion, and in the controversy that followed, few on either side heeded Dr Pusey's comment that only 'un-science, not science, is adverse to faith'.

That, in brief, is the history of our swifts' nesting place. We chose to study this species, first because we sought to know more about the most aerial of birds. Secondly, its habits were largely unknown. Thirdly, swifts breed late, with eggs in June and young in July, so that their study could be combined with that of birds which nest in April. Finally, a little before our arrival in Oxford, a schoolboy had brought to the zoology department the meal carried by a parent swift to its young, which had been ejected when he caught the bird to put a ring on its leg. He had found an accessible nesting colony in a nearby village, and it was here that we first went to watch swifts.

The scene was quiet and picturesque, with half-timbered cottages in red brick, thatched roofs, and an ancient church with a sundial instead of a clock. Closer knowledge showed the underlying poverty, with holes in the plaster-work on the walls and in the thatch of the houses; and a sundial was useless in the wet summers of the first post-war years. But because there were holes in the thatch and under the roofs, swifts were nesting, some of them only ten feet above the ground, which is unusually low.

Here we studied what we could. By setting up a ladder outside the houses, we could reach an arm into the holes and so could bring out the eggs, young or adults for inspection or weighing, to the pleasure of the small children who attached themselves to our party whenever school hours permitted. But it was not a good method. Some of the nests were too far inside to be reached, while some of the adults deserted because we caught them, though we did not realise this until later. We were able to measure the growth of the young, and we discovered their astonishing ability in cold weather to survive long periods of cooling without harm. But we could not see what was happening in the nests. Surely we could do better?

There were, of course, swifts nesting in the city of Oxford itself. But bedroom curtains were fiercely pulled across when at dusk, I gazed up under the eaves of lodging houses in Wellington Square. In any case, these nests were inaccessible. An unfounded rumour that the birds were nesting in the belfry of a college chapel led to an entertaining, though fruitless, clamber among the bells. Some pairs nesting over a hundred feet above the ground in the University Museum we dismissed as hopeless for study.

But in September 1946, I visited Switzerland and heard of E. Weitnauer, the schoolmaster of Oltingen, a village in the Jura mountains. Swifts nested in the church there, and Weitnauer had the idea of making further nesting places for them under the eaves of his house. Swifts later used these nesting boxes and he began a study that was to last many years. At that time, none of his observations were published, but a long paper appeared in 1947. It was this visit that provided the basic idea for the experiment at Oxford.

In the Museum tower, hereafter simply called the tower, swifts were nesting in the ventilator holes, of which there were ten on each of the four sides of the tower (Plate 5). The ventilator shafts had been fitted when the tower was built, but as they ventilated emptiness, and as pigeons entered through them to nest inside the tower, they were afterwards sealed at both ends by wooden blocks. At a later date, workmen broke holes in the wooden blocks across the ventilators in order to pull up slates for repairing the roof. They replaced the inside blocks but left the holes in the outside ones. Swifts discovered these holes and entered them to nest inside the ventilators on the broken wooden ledges; and they had been nesting there for many years when we first saw them.

Our plan was this. The tower, as already mentioned, rose about sixty feet from its base in the upper part of the building. First, a wooden platform was erected inside the tower about half-way up, and four feet below the level of the lowest row of ventilators. This platform had a large gap in the centre to allow light to enter from the only window in the tower, which was below it, and it was reached by a thirty-foot ladder. A second and smarter platform was then built above the first, and four feet below the level of the third row of ventilators. This was reached by a fixed ladder from the lower platform. These platforms gave easy access to two of the four rows of ventilators, and the other two rows were reached by portable ladders from the platforms. We then, and here lay the risk, removed the ventilator shafts and substituted nesting boxes projecting horizontally inwards. The swifts could enter these boxes through the same holes which formerly led into the ventilators, hence from outside nothing was changed. But, inside, the birds would now have to proceed horizontally instead of upwards to reach suitable nesting places. Actually, in case this change should be too disturbing, we at first left the uppermost row of ventilators unmodified.

These changes were completed in April 1948, before the swifts returned to Britain from Africa, and on one of the first days in May we waited anxiously outside the tower at dusk to see what would happen. A swift arrived, circled the tower several times, as is customary, and then entered one of the holes. We waited further and to our relief it did not again emerge, but stayed for the night. In the next fortnight many others returned, and so far as we could see, all of them accepted their modified homes in which to roost. We then converted the remaining ventilators. Soon afterwards the birds built nests and laid eggs in the boxes and we started our detailed observations.

Each box was fitted with a lid near the back, which we could lift, and in this way we could bring out the eggs or young as we wanted. Most of the adult swifts showed little or no fear of us. This is presumably because, in nature, enemies cannot reach their nests from the back, and so the birds are not adapted to deal with such a situation. If a hand was inserted into the box, a few of the adults showed mild alarm, but many ignored it, while others clawed it vigorously, in one instance so effectively as to set up blood-poisoning. We found that we could even place bird-rings on the legs of some of the adults in the boxes without their minding, but if we lifted them out for ringing or weighing some of them deserted. This was unexpected, especially in view of their tameness in other respects. Once we realised that it might happen, we stopped ringing the birds, though this unfortunately meant that we could not recognise each bird individually, which would have helped some aspects of our work.

Even now, we could not actually watch the birds. So in the following year we removed the wooden backs of the boxes and replaced them by glass. We could now sit in darkness on the upper platform, watching the birds by means of the light from the entrance holes, and could see what they did from a distance of a few inches

Fig 2: Inside the tower

without their being aware of us. It was fascinating to observe them in this way and to realise that we were watching behaviour which, except perhaps by our Swiss colleague, had not been seen by man before. We sat on a stool of such a height that our eyes were on a level with the second row of boxes, and to help in recording, we took an old petrol drum, cut out one section of it, and placed inside a small electric battery and bulb and a writing board. With the open section of the drum towards the body, and a notebook inside, we could sit facing the bird and could record in the light without the bird being able to see it, as shown in Fig. 2.

On the lower platform, as already mentioned, light came from below through the central gap, so here we fitted curtains (the remains of war-time black-out) over the glass. These lower boxes were used chiefly for routine work, all detailed observations being made on the upper platform. In 1954, however, we planned to photograph the birds by means of the new technique of electronic flash. This might have caused disturbance on the confined upper platform, so the lower platform was used. The central gap was covered over, to reduce the light from below, and three of the nesting boxes were fitted with a large sheet of glass at the side (instead of a small one at the back). A camera was then set up on a tripod on this side of the box. A difficulty that we have not yet mentioned is that adult swifts tend to breathe on the glass back of the box and to dirty it, and when we are watching we have to clean the glass regularly. For photography, extremely clean glass was required, so each time before photographs were to be taken, the glass side was slid out and replaced by a thoroughly clean one. The design of the boxes for photography and all the attendant arrangements were carried out by Mr H. N. Southern, and his unique photographs provide the main illustrations of this book (see Note on page 11). It may be added that swifts usually show fear of a strong light, but the pair that were most photographed became so used to it that we could later shine a torch full on them without at all disturbing them.

From 1948 onwards, many bird-watchers and zoologists have visited the tower. We start our tour on the ground outside, pointing out the swifts wheeling and screaming in the air a hundred feet above. This we have taken care to do since the day when we showed the birds in the boxes to a distinguished laboratory zoologist, who was surprisingly unimpressed; we later found that he had not realised that he was looking at truly wild birds, a tribute to their tameness and his ignorance. Entering the museum, we proceed by a staircase to the first floor, pausing briefly before the room, now marked by a plaque, where Samuel Wilberforce and T. H. Huxley held their famous debate, and perhaps admiring the fragments of the last dodo. To reach the base of the tower, we then ascend by a steep, narrow, spiral staircase, the spiral, though built in the eighteen-fifties, being correctly arranged so that a defender in the tower above would have his sword-arm clear. To complete the medieval atmosphere, this staircase is for part of the way in complete darkness, though higher up there are slits for the use of defending bowmen. We then enter the tower, by a locked door which, however, we keep open when anyone is inside, in case they should fall and need help. From the floor of the tower, a bare ladder without a handrail reaches for thirty feet to the first platform, and at this point various zoologists have found themselves unable to proceed further, though we ourselves have got so used to it that we ascend and descend freely, even in the dark or carrying bulky equipment. It is, however, less embarrassing when a visitor gives up at the foot of the ladder instead of half-way up, as has occasionally happened; one zoologist lost heart just before he reached the first platform, when his head but not his feet had reached safety. But so far we have had no casualties.

After a brief rest at the first platform, we make the much simpler vertical climb to the upper platform, and here, if we are lucky, the birds will be active around us. This depends somewhat on the weather, as on a cool day when feeding is difficult the birds stay out most of the time. For those birdwatchers, and they are the majority, who have hitherto seen swifts only in flight, the first view at close quarters is a surprise. 'Why,' exclaimed a distinguished naturalist, 'they look like mice.' The remark reflects the fact that though the swift measures some 16 inches across the wings, it is only about 4 inches from the tip of the beak to the base of the tail, and it weighs only about

1½ ounces. Further, though so agile on the wing, it crouches clumsily when beside its nest.

On this second platform, close under the slates, we have spent many hours. Sitting in the darkness, very cold on a cold day and extremely hot on a hot one, we have not envied our fellow bird-watchers in their more conventional attitudes on marsh or mudflat. The passage of time is marked by the clock of Keble College over the way, always a little slow and a little out of tune, but otherwise we are cut off from the outside world as, undistracted, we seek to study, so far as a human being can, the way of life of the swifts.

Chapter 3

SETTLING IN, AND THE FIGHT FOR HOMES

In our first season at the tower, in 1948, sixteen pairs of swifts laid eggs and several other pairs occupied boxes but without laying. From then on, the number of birds has gradually risen, reaching thirty pairs with eggs and six non-breeding pairs in 1955. This is actually more than the tower held before our experiment began, since the bottom row of ten ventilators was at that time blocked and only thirty were available for swifts.

The first swifts appear over Oxford in the last days of April. At this season the weather is usually cool and insects are scarce, so that the birds spend nearly the whole day away feeding and are not seen in or around the tower except at dusk, when they come in to roost. From early May onwards, more birds gradually arrive, until by about the fourth week in May the colony has effectively assembled. Ringing has shown that the adult swift normally returns to its nesting hole of the previous year. Young birds select a site as yearlings in their first summer, form a pair and build a nest, occupying the site throughout the summer but without laying eggs. They then return to the same site in the following year to breed.

The swift usually enters its hole by a straight flight with a short rise at the end, alighting on the rim of the hole and immediately running in. The fly-in may be for thirty to fifty yards, and accurate judgement is needed to bring the bird at high speed to the correct spot (Plates 15, 16). If it misses the entrance, as occasionally happens, it usually clings for a moment to the wall, but it does not then try to enter its hole by walking sideways. Instead, it drops off and tries another flight in. With a strong or gusty wind, and also at dusk in fading light, the birds are less accurate than at other times, and misses are frequent. This was particularly so for the birds nesting in the thatched roofs of the village, which had a less clear fly-in than at the tower, and here on one occasion we saw a bird make twenty consecutive attempts to enter its nest-hole at dusk, eventually giving up and flying off. That these misses at dusk are due to the bad light was shown by the Swiss worker Weitnauer, who, after a bird had failed several times to enter a hole at dusk, shone a bright torch at the eaves, whereupon the bird entered successfully.

In the first few days of May, the colony is often a silent place, but so soon as there is warm weather, parties may be seen screaming round the tower, and the birds regularly visit their boxes by day. The most favoured times for such visits are for an hour or two in the morning around 7.30 a.m. and again in the late afternoon around 6 p.m. The time when the birds are least likely to be there is between noon and 4 p.m., when airborne insects are usually most numerous. Before breeding starts the birds may spend only a few minutes at a time in the box and then leave for another flight round; at other times they stay for an hour or two, even in fine weather, perhaps for a rest, and they regularly come in for shelter during rain. Indeed, in continuous heavy rain the birds may stay in their boxes almost the whole day. Thus on 26 June 1951, we watched at the tower for ten hours (8 a.m. to 6 p.m.) in cold, windy weather with almost continuous rain, and a pair without eggs or young spent respectively only 2 and 12 (out of 600) minutes outside the tower. (The pairs with young had, of course, to go out to seek for food.) When in its box, a swift, which has very short legs, normally rests with its body on the floor, raising itself on its feet only when moving about.

Each box is occupied by only one pair, and they defend it against all other adults. If strange individuals are flying about outside the colony, the owners of nests often come in and then sit looking out of their entrance holes, their pale throats very prominent (see Plate 3), screaming violently as the potential intruders pass by. If both owners have returned, they usually sit side by side in the entrance and scream in duet, one giving a higher note than the other. This, we think, must be the 'swee-ree' call of the swift, mentioned in the *Handbook of British Birds* and other standard works; it is really two notes, one given by each member of the pair. Such 'duetting' is uncommon in birds, but has been recorded for various tropical species. Thus in the bokmakierie, a South African shrike, the first bird calls 'bok-mak' and the second 'kierie', so as to make one continuous call. Likewise in Australia the male coachwhip bird calls like the crack of a whip, while the female immediately follows with a soft 'gee-up', and it was a long time before naturalists realised that the call was a combined effort by the pair.

If, as occasionally happens, a strange swift enters an occupied box, the owner screams, rises up on its feet and advances towards the intruder with wings held partly out and raised. Often, it raises especially the wing nearest to the intruder, tipping its body sideways and exposing the legs and feet, which will play a prominent part if a fight develops. This behaviour is a typical threat display and I have seen a similar display in the large alpine swift in Switzerland. Swifts, like so many other birds, use psychological warfare as a prelude to true fighting. Often, as in other birds, the threat display suffices, and the intruder quickly leaves the box and does not return. The threat of the owner must, to be effective, have a counterpart in the timidity of the intruder. Even when the owners are absent, a strange swift entering a box acts nervously. It sometimes stays quietly in the entrance hole before proceeding further in, and then explores tentatively, perhaps flicking its wings or walking high on its feet, and it may leave, of its own accord after a minute or two. Most strangers are probably looking for a vacant home and merely seek to know whether a box is occupied.

Occasionally, an intruding bird stays in the box even if the owners are present, and then a serious fight may occur. We have seen about twenty such fights, all except one of them in May or early June before the eggs were laid. It is at this season that newcomers must find nesting sites if they are to breed, and the only fight seen later in the year was probably due to a bird entering the wrong box by mistake. Six of the observed fights started in the morning, one at midday, three around 5 p.m., one at 8 p.m. and two (brief ones) at dusk. The others were first noticed on our routine visit in the evening and we do not know when they started. Omitting some very brief scuffles, the shortest of such fights lasted for 20 minutes, the longest for 5¾ hours, while we found another fight already in progress which continued for 5½ hours. In one box there were fights on two consecutive days, but we do not know if the same intruder was involved in both, and at another nest a fight took place three days after an earlier one.

The fight is usually preceded by a few seconds of excited screaming and posturing. The birds then rush together, tipping the body sideways and away from the intruder so as to have their legs clear, then gripping each other's legs with their claws and struggling furiously. They remain thus gripped for the rest of the time, that is, for up to five hours. The claws are sharp and their grip is extremely strong (they easily draw blood from the human finger), but as the birds usually grip their opponent's legs, which are hard and scaly, little if any damage results. The birds also peck repeatedly with their beaks at whatever part of their opponent's body comes within range, usually the feathers of the body or wing. The pecks are vigorous but, because the bird has a soft beak, they are harmless, as we tested by inserting a finger during a fight and receiving some of the pecks on that. During the fight, the birds also struggle with their wings, pressing on the floor of the box and heaving about, sometimes shifting their relative positions. They also scream loudly. But such periods of violent activity alternate with pauses when the birds lie motionless and silent, apparently exhausted, with the nictitating membranes drawn over the eyes, presumably for protection (see Plate 2).

After a time, one of the fighters will be found lying on its back below the other (but with feet still interlocked). Surprisingly, it is usually the under bird which is winning. In this position it can shift its opponent, and it gradually moves it towards the entrance hole, the other resisting, and though it is hard to be sure of what is happening as the birds are closely grappled, the under bird appears to be trying to throw its opponent out of the hole. At this stage, the upper bird has sometimes tried to escape on its own, but it cannot do so as the victor does not relax its grip. Eventually, the fight carries on over the hole itself, with one bird partly outside, and the usual end is for both birds to tumble out, still grappled, though occasionally one has managed to remain within. One fight continued with one of the combatants dangling half out of the hole, flapping hard, for as long as twelve minutes.

In the later stages of a fight, the bird that appears to be getting the worst of it utters a plaintive piping call, not heard in any other circumstances in the life of the swift. It appears to be in great agony, alternately uttering this call and lying back breathing faintly with eyes closed, as if exhausted, while the other continues to peck it. We found it a painful proceeding to watch, a gladiatorial show conducted only a few inches from our eyes. But in fact the beaten bird is usually little if at all hurt. Thus on one occasion we accidentally disturbed two birds after they had been fighting for 4½ hours. The presumptive victor thereupon ran to the entrance, though it did not leave, while the other lay back in apparent exhaustion. After about four minutes, the victor returned, sat beside the other and pecked it. There was no response, its opponent seemed half-dead, so it got onto it and started dragging the unresisting body to the entrance. At this point, we again frightened it and it let go its hold and left the box. A moment later, the apparent corpse rose up and also left the box, evidently uninjured, though for several minutes it had looked to be dead.

Most other species of birds use threat postures in their fighting to a much greater extent than swifts. One could scarcely expect elaborate displays to be evolved by birds which fight in such cramped and dark quarters. But even though swifts usually fight by more direct means, their fighting resembles that of other birds in that the combatants are rarely injured.

Since we were watching inside the tower, we could not see what happens when the grappled pair tumble from the nesting hole. It is clear that usually they soon separate, since after a short interval, the owning bird returns alone to the box, and it then spends a long time preening itself, perhaps half an hour or more. There are, however, two published records of swifts found alive but interlocked on the ground below their nests, presumably unable to release their grip. Such birds would be likely to fall victims to a cat or other enemy. Hence fights may sometimes prove fatal. Once also, on 20 May, we found in a nesting box a freshly dead male which was possibly killed in a fight. Two other swifts were in the box at the time, but we did not see what led up to this situation.

In every fight, one member of the pair owning the box took a far more active part than the other. As our birds were not marked, we do not know for certain whether the active bird was the male. The more retiring bird sometimes took part for a short while at the start of a fight, screaming and attacking the intruder, but after a few minutes it usually retired to the nest at the back and took no further notice. Sometimes it left quietly and stayed out, presumably feeding. One such bird calmly continued building the nest, though its mate was furiously grappling with a rival only a few inches away. In another fight, the second owner entered and left the box several times, apparently paying no attention to the fight. In another instance the second owner even preened the throat and nape of one of the fighting birds, presumably its mate, in a typical courtship action! One fight was unusual in that four individuals, evidently two pairs, were involved. Two birds were fighting hard with a third (presumably the other owner) joining in occasionally. Then a fourth swift entered the box. The third bird at once left the other two and attacked the newcomer, which after five minutes was driven out. The third bird then rejoined the original fight. After twenty minutes, one of the three left or got pushed out, but the other two continued to fight for some time.

An exceptional fight occurred on 13 July 1950 in a box with a well-grown nestling. Soon after 11 a.m. an adult entered and fed the chick. Six minutes later another adult entered, its throat-pouch full of food. The parent usually pays little attention to its mate entering with food for the young, but this bird was at once attacked. After a fight which lasted, in all, for fifteen minutes, both birds fell out of the hole in the usual way, and a few minutes later one of them returned without food and preened itself. This is the only fight which we saw with young in the nest, and it was presumably due to an adult entering the wrong box by mistake. That adults may make such mistakes is also shown by their occasionally coming with food into an empty box, their own young presumably being in the adjoining one.

We suppose that the spring fights are primarily for the ownership of a suitable nesting hole and not for mates. Almost certainly, on the occasion when four individuals were involved, the fight was not to obtain a mate. Usually, so far as we know, the fight started when a strange bird entered the box. In one instance, however (and another has been observed in Switzerland), three birds entered a box one after the other and then fought, perhaps having started their battle in the air outside, though this is not certain.

As already mentioned, both the adult swifts and the former yearlings normally return to their nesting sites of the previous year. Hence the fights in late May are perhaps due chiefly to birds which, for some reason, have found their former nesting site destroyed or no longer available and so are forced to settle elsewhere. Later in the summer a number of new pairs appear in the tower and occupy empty boxes. Some stay for only a few days but others for the rest of the summer, returning next year. Most of these newcomers are probably yearlings. They evidently seek unoccupied boxes, and do not, so far as we know, try to dispossess established adults, which would probably win in a fight.

Probably connected with the discovery of empty nesting places is a curious type of aerial display which we have called 'banging'. Often when we were sitting in the tower we would hear a sharp bang on the outside of a box. At times both members of the occupying pair were at home, so we knew that the bang was caused by a stranger, and this was confirmed by the excited screaming of the owners. We could see more of this behaviour by watching outside. A solitary swift circles the tower in leisurely flight, then goes up to a nest-hole and brushes or bangs against it and flies on. Sometimes several swifts fly up in follow-my-leader style, each brushing the box as it passes. A lone bird or a party may thus fly up to the same box several times, or they may select different boxes in succession. The banging bird is usually silent and it flies unhurriedly. Usually, it just touches the entrance with its wing and then flies on, often it comes up and passes on without touching it at all, while occasionally it alights and looks in. Rarely, it enters. It may not get beyond the entrance before leaving again, and if it intrudes further it acts very nervously. If there are young in the nest, it ignores them, even though they may beg for food. If one of the adult occupants of the box is present, it of course attacks the intruder at once. Moreover, if bangers are flying around, many of the birds owning boxes enter them and sit in their holes screaming as the bangers fly past. The whole behaviour suggests that strange birds are seeking unoccupied holes and that owners are proclaiming their tenancy, and the habit probably helps newcomers to discover which holes are occupied without the risk of a fight.

Banging occurs only in good weather with little wind, and is specially common on the first fine day after a spell of bad weather. It is commonest around 8 a.m. and infrequent in the afternoon, though it may occur at any time of day. It occurs throughout the breeding season, and, as already mentioned, new pairs may come to occupy a box at any time in the summer. I have seen similar behaviour in the Galapagos shearwater. In the afternoon a small group of these birds would often circle over the water close to the cliffs where others were nesting, at times flying up to a nesting hole, at which the owner within would call and the party would fly on. Fulmar petrels in this country perhaps do the same sort of thing, but whether their behaviour has the same significance as in swifts I do not know.

Swifts compete for holes not only among themselves but also with other species, the chief enemy being the starling. At the tower, a pair of starlings which had built in one of the boxes in April was successfully turned out when the swifts returned in early May, but we did not see them do it. To reduce the number of such conflicts, we now block up the holes in the tower until the end of April, by which date most of the local starlings have already started nesting elsewhere. Even so, a few sometimes get in before the swifts return. In one such instance, a single starling had taken over a box. On the day that the swifts returned, they came in at dusk before the starling. When the starling came up to the entrance, one of the swifts screamed and raised its wings and after a few moments the starling left and did not come back that night. On the following evening, the starling got in first. A few minutes later the swifts came in, almost together. They stayed still in the entrance, the starling stayed still at the back of the box, and all three 'froze' for about five minutes. One of the swifts then gave a feeble threat display, but did not follow it up. It was now getting very dark and, not liking to leave the birds together for the night, I gently raised the lid of the box, intending to catch the starling. The slight noise broke the spell. One of the swifts advanced in a typical threat display, high on its toes, wings arched and body tilted sideways, screaming hard and at intervals thumping with its wings on the floor of the box. The starling stayed still, presumably because it does not understand the swift's methods of expression. Because it stayed still, the swift in turn seemed nonplussed and stayed still. After another pause, the starling made two rather feeble jabs at the swift's head with its beak, the swift then rushed in under its guard and gripped its body, there was a flurry and scuffle, and both moved rapidly to the entrance, passing over the body of the other swift and out. I heard the starling give a final squawk as it flew off. It was too dark for the swift to return, but its mate stayed in the box and on the next day both were back. The starling did not appear again.

Published records show that, while swifts are sometimes successful in displacing starlings, victory sometimes goes the other way. Often the dispute does not result in injury, but when, as sometimes happens, the two birds fall to the ground, the starling has sometimes been able to wound or kill the swift with its beak, while a starling has been found with its breast torn and bleeding, probably from the claws of a swift. In another instance where both birds fell to the ground, a cat caught the swift before it could take off again.

Swifts at times displace house sparrows from holes in thatch, in which one would suppose they have little difficulty, though there is a published record in which house sparrows apparently broke a swift's eggs. Hole-nesting birds are in general safe from egg-eating mammals or birds, whereas those that nest in bushes or on the ground are often robbed by weasels, jays and the like. But suitable holes are usually too few to go round, so that competition for sites between hole-nesting species may be severe. Their added safety from predators is obtained only at a price.

The high-pitched scream of the swift, heard in threat display, in social screaming-parties and between the pair, sounds to our ears simple, harsh and monotonous, but this impression may be false. In 1955 a microphone was placed alongside the nesting boxes in the tower and the swift's noises were transmitted by cable to a BBC recording van in the museum drive. When the recording of the calls was later played back at a quarter of the natural speed, it revealed unexpected diversity. The scream usually starts with a group of separate notes, then becomes a nearly continuous bubbling call, which gets louder and rises somewhat in pitch, then falls again, and ends with several separate notes. At quarter-speed, it sounds like the thrilling vibrated cry of the great northern diver (or loon), while when slowed down to a tenth of the natural speed, it is like the clucking and crooning of a domestic fowl. Other notes, and modifications of the scream, were also revealed. It seems reasonable to suggest that the swift itself can hear at least much of this diversity.

The songs of many woodland and garden birds are also richer and more musical than is apparent to us, and it has now been proved that at least one such species hears more of them than we do. The whip-poor-will, an American nightjar, has to our ears a song of three (or at most four) notes, corresponding with its name, but a recording of the song played at half-speed showed that it really consists of five notes. The recording of a mocking-bird imitating the song of a whip-poor-will likewise appeared to consist of three notes at normal speed, but when played at half-speed revealed all five, showing that the mocking-bird heard and reproduced the song correctly.

Played at half-speed, the recording of the whip-poor-will sounded like the song of its larger southern relative, the chuck-will's-widow. Actually it was when I read this that I determined to get a recording of the swift's scream, because the larger alpine swift of southern Europe has a call which to our ears sounds utterly different, being a musical trill. It seemed unlikely that related species would differ so markedly as this, and in fact the recording at quarter-speed showed the note of the common swift to be of a similar general pattern to that of the alpine. If we could hear properly at high frequencies, we would probably think our swift's shrill scream beautiful.

Chapter 4

COURTSHIP

The adult swift normally forms a pair with the same mate as in the previous year. The Swiss recoveries of ringed birds make this certain, and our own few records (before we gave up ringing) are in agreement. This does not mean that swifts pair for life, as the pair probably separate each autumn and rejoin in the spring. If they stayed together for the winter, they would presumably migrate together, but this is not so. Thus in four years we checked the return in spring of all the adult swifts to the tower, by visiting it each evening as the birds came in to roost. In just over three-quarters of the pairs, the two individuals returned on a different day, the interval varying between one and twenty-one days. We also checked the departure of the adults in four autumns, and found likewise that in just over three-quarters of the pairs the two individuals left on different days. In a quarter of the pairs, admittedly, both birds arrived or left on the same day, but this can reasonably be ascribed to chance, and there is no need to suppose that they kept company on migration. Probably, the same individuals pair together in successive springs because they return to the same nesting hole.

In small song-birds, it is unusual for the same cock and hen to mate together in successive years, but this occurs not infrequently in various larger birds, such as the crow tribe, shearwaters and penguins. In some species of this type, as in swifts, the two birds probably separate for the winter and meet at the same nesting place in spring, but in others, such as crows, the pair stay together for most of the year (though they sometimes change their mates).

On the first night in May in which the two swifts of a pair

occupy their box together, they often settle down peaceably at once, with much mutual preening. Occasionally, however, there is some friction. Thus in one box, occupied by a single bird on the previous night, the first to come in greeted the second bird as it arrived with much threat display, screaming and partly raising its wings. The second bird did the same and there was then a short scuffle, though the birds did not actually grip each other with their claws. After that, they quickly settled down together for the night. In a similar instance in another box, a scuffle was followed by one bird leaving, but it flew in again two minutes later, after which there was a second scuffle and the pair then settled down together.

The pairs concerned in these observations were probably meeting again after the winter. It is to be expected that pair-formation for the first time would take longer. We perhaps saw one such instance, but as the birds were not marked, we cannot be sure. A single bird was in occupation of a box, its mate of the previous year perhaps having died. By 24 May, when almost the whole colony had returned, it was still unmated. At dusk that evening, when the occupant was already in for the night, a strange swift came up to the entrance, screamed rather mildly and held its head up, exposing the white throat. It stayed in the entrance and the original occupant advanced half-way along the box towards it, screaming hard, and also holding its head up. Both then stayed still, a few inches apart, for five minutes. The newcomer then came in a little further, paused, and then advanced again, keeping its head well up in a strained position, as did the original occupant. The latter then touched the newcomer's throat with its bill. Fifteen minutes later, when it was getting too dark to see more, the birds were still in about the same positions, and both evidently stayed in the box for the night. The following night only one bird, presumably the original one, spent the night there. No more observations were made for six days, when soon after noon a pair entered the box one after the other with much screaming, and with partly tilted and half-raised wings, as if about to fight, but one kept its head up in the same strained attitude as before and the other then preened its throat. The pair then settled down together in a bout of mutual preening.

These observations, together with those described in the last chapter, suggest that the first reaction of an unmated swift to a newcomer is hostile. If the newcomer is merely seeking an unoccupied nest-hole it retreats, or occasionally it may stay to fight, but if it is a potential mate it either retreats and quickly returns, or stays still, stretching up its head and beak and exposing its pale throat (see Plate 3), which may be interpreted as a submissive posture. In the robin and many other birds, it is now well established that the first stage in pair-formation looks very like a fight, with threat display and song. The newly arrived hen distinguishes herself from a rival by repeated return to the cock and, in many species, by a submissive display, and gradually the two birds become accustomed to each other, though it may be several hours or even days before their threat display subsides. 'The Taming of the Shrew' is acted each year in the pairing of birds. It is a pity that we could not establish the story more certainly for our swifts with the help of ringed individuals.

An unusual incident occurred in May 1955, when both of two adjoining boxes were regularly occupied by single birds, while a third bird alternated between the two. It was first seen on 7 May and spent six nights in the same box. Then it moved next door for a night, but came back to the original box on the following night. The next three nights were spent next door, it then returned for three more nights, went next door for a night, back for a night, and finally moved next door and stayed there to breed. It seemed equally at home in both boxes and courted freely with both owners. The sex of this bird was not known. In Switzerland, however, Weitnauer was able in three instances to show by ringing that it was the male that first held a box alone. He also saw an unusual incident in which a female swift took the initiative in pair-formation. This bird lost its mate shortly before the time of egg-laying. On the following day it flew slowly in front of the colony with vibrated wings and a long-drawn call, a male following behind. Five times it flew up to and entered its box, the male each time flying on, but on the sixth occasion the male also entered and coition followed.

Sometimes a new pair have arrived together at the tower to take over a box. Presumably they had formed into a pair elsewhere before moving. This was definitely the case with a Swiss pair which bred one year in a nest-box on a tree that was later removed. On their return in the next spring the pair repeatedly flew up to where the box had been. Later they moved and bred together in the near-by church tower. An English observer has also seen a pair of swifts repeatedly fly up to a blocked-up hole in which swifts had bred in the previous year. Swifts, sometimes single and sometimes paired, arrive at the tower at any time during the summer, most of these presumably being yearlings that are forming pairs for the first time. Some of them move on after a few days, but others stay for the rest of the summer and return in the following year. Single individuals rarely stay unmated for more than a few days.

Once two swifts have formed a pair, they recognise each other individually. Thus when a strange individual enters an occupied box it is at once attacked, whereas the mate is not. The mate as it enters is greeted with a short scream, perhaps with partly raised wings, and the entering bird greets the occupant in the same way. For a long time we called this a 'greeting ceremony', and it was not until we knew swifts well that we realised that it was similar to a mild threat display. Perhaps, therefore, it really is a threat display, which dies away as soon as the birds have fully recognised each other. Supporting this idea, the display tends to be less excited later in the season, as the pair become used to each other. The existence of potential hostility between the pair is further shown by the fact that, just after an intruder has been thrown out in a fight, the two members of the pair show increased threat behaviour towards each other. Thus on an evening in mid-May when the migrants were still arriving, a third swift tried to enter a box where two were already present. After a brief fight it was ejected. The owning pair then had a brief scuffle together before again settling down side by side. A few minutes later the stranger again entered, it was again ejected, and there was again a scuffle between the pair, after which tension remained so high that they settled down for the night on opposite sides of the box, instead of side by side as is normal.

There is similar evidence in the robin of tension when cock and hen come close together, and here also it is intensified just after a fight with an intruder. The same is found in other birds. Probably, therefore, a true threat display is involved when the pair meets. On the other hand, at least in the swift, the performance is so regular and so ritualised that it has all the appearance of a greeting, and one wonders whether it might not in fact be one, though originating from threat display. This raises a problem of behaviour which cannot be solved without a parallel study of other birds, and for that matter of other animals also. Why, for instance, do men and women greet each other by shaking hands or embracing? These actions have some resemblance to fighting, though the gripping is of a different kind. Again, the smile that accompanies a greeting might recall a snarl, though it does not expose the canine teeth, as primitive man perhaps did when fighting; and we do not normally mistake the friendly intent of a smile or embrace.

After the initial threat (or greeting), the arriving swift usually comes quickly up to its mate, which often advances to meet it. Then, as already mentioned, one or both birds may hold up the head and beak, exposing the throat, probably as a submissive posture. Submissive postures are found in many other birds, though usually with the head held low and beak pointing downwards, whereas the swift holds head and beak up. But most birds attack with the head high, swifts with the head low, so that swifts agree with other birds in that their submissive posture is the opposite of that which precedes attack. As in the submissive attitude of other birds, a vital part of the body is exposed to potential attack, the throat in swifts and the back of the head in many other birds. Men, likewise, bow or bare the head when greeting their friends.

Almost immediately after a pair of swifts have met in the box, they start to preen each other, often first on the exposed throat. This mutual preening is the chief courtship action of the species. It occurs throughout the breeding season, being most vigorous early on, before the eggs are laid. The bird preens mainly those parts of its mate's body which it cannot reach for itself, the throat, the nape and other parts of the head. I have watched similar mutual preening in the alpine swift in Switzerland, though it seemed less frequent than in our own species. Mutual preening is also found in pigeons, parrots and various sea birds such as gannets.

A swift spends much of its time in the box preening itself, because for rapid flight it is essential to have the surface feathers as smooth as possible. However, the mutual preening of a pair of swifts seems to be much less of practical than of ritual significance. Similarly, a cock robin feeds its mate in courtship, and though the food is eaten by the hen, the ritual seems the important part of this behaviour. There are many instances of such 'derived' activities in the courtship of other birds. The mutual preening of swifts, like the courtship-feeding of robins, is not connected with coition, but occurs throughout the summer. It perhaps helps to inhibit potential hostility between the pair, or to cement the pair-bond, and so may be compared with the affectionate behaviour of man and wife. Both the initial threat-like display and the mutual preening of the pair become much less excited in the latter part of the breeding season, in the same way that the delighted greeting of the newly-wed gives place to the casual

'Hullo' of later and more contented years.

Mutual preening is sometimes accompanied by quivering of the wings, and when they are very excited the birds sit close together in a rather humped position with the feathers fluffed out. One of them also gives a call, which is much softer and of lower pitch than the usual scream. This call is characteristic of the courtship period and is often heard apart from mutual preening, but it ceases when the eggs have been laid. It is uttered by only one of the sexes, we do not certainly know which, but it is probably the female, as the call is sometimes heard in flight-chases from the leading bird, which is usually the hen. The call is given chiefly when the pair is together in the box, occasionally when one is there alone. If the birds lose their eggs or young, the call is resumed, and we have also heard it from parents after their young have flown in August. The yearlings and other pairs without eggs or young use this call throughout the breeding season.

Coition occurs in the boxes, usually around either 7 a.m. or 6 p.m. (GMT). The female sits in the usual resting position, the male mounts, gripping the female's back with his claws and her nape with his beak, the female raises the tail and the male twists his body round to make contact. Usually, the male mounts three or four times in succession, but sometimes only once. Afterwards, there is often mutual preening, and beforehand a characteristic subdued scream. Coition is confined to the period just before and during the laying of the clutch, and as in the robin and many other birds, it is attended with little apparent excitement and scarcely any prior or subsequent display.

There has been great dispute as to whether or not swifts also mate on the wing. Only a few observers have recorded this, and swifts are the only groups of birds credited with the habit, hence caution is needed, especially as it is hard to be sure of what is happening above one in the air in birds in which the two sexes look alike. Nevertheless I was convinced of its occurrence by the published evidence, and later saw it unmistakably myself.

Aerial display often starts with one of a pair flying in front of the other and suddenly holding its wings out stiffly high above the horizontal, sometimes almost vertically. It then falls forwards and downwards, at which the second bird gives rapid chase. Often the first then changes to a quivering flight, in which the wings seem vibrated rather than beaten. After a few moments both birds resume a normal but slow flight, one close behind the other. The one that is behind then comes above the other and alights gently on its back, sometimes making one or two unsuccessful attempts, which are perhaps part of the ritual, before getting hold. The pair then descends in a very shallow glide, the female holding the wings out horizontally and the male holding them upwards at an angle, while both spread and twist their tails. During the glide both birds may hold their wings still, but sometimes one of the pair beats them rapidly in a small arc, and if they lose too much height both may do so. After a few seconds, they separate.

This is a composite description, based on my own observations combined with those of others. I have seen the downward glide with raised wings followed by flight with vibrated wings and this in turn by slow flight with one close behind the other. Often, nothing further happens. I have also seen the characteristic slow flight followed by aerial coition, in which, after the male had twice bumped the female's back but came off again, he got hold and the pair descended in a shallow glide, the male beating its wings, the female gliding, and both spreading their tails. Other observers have seen the quivering flight lead on to aerial mating, while Weitnauer has seen it lead on to coition in the nesting hole (see p. 41). Sometimes, but not usually, the leading bird in a chase gives the soft call already mentioned, and several observers have recorded a characteristic scream during aerial mating itself, though in one watched by me the birds were probably silent and two other observers were definite that the birds were silent, hence the scream is evidently not essential to the performance. As already mentioned, coition in the boxes is often, though not always, accompanied by a soft scream.

The birds that I watched met 60 feet above the ground, and in other accounts the height has been reported as between 40 and 80 feet, once as high as 200 feet and once as low as 30 feet, though not at the start of the glide. The downward glide is usually shallow but sometimes the birds sink more rapidly. Once, excited screaming drew my eyes to two swifts, one on the back of the other, falling almost vertically, but with wings held out so that they spun round in a horizontal plane. Perhaps they had got out of balance. Most of the published records, and the two seen by me, occurred around either 7 a.m. or, less commonly, 6 p.m., the two times of day when mating is most frequent in the boxes.

Coition at the nest was first recorded in 1906, in the Protestant Church at Wolfstein in Germany. Aerial mating, on the other hand, was described by Gilbert White at the end of the eighteenth century, but there were few other records of it until the nineteen-thirties, when correspondents to a German ornithological journal brought further instances to light. The reason that so few observers have seen it is that the birds do not draw attention to themselves and hold contact for only a few seconds. If they made themselves conspicuous, they might well be exposed to attack by falcons. That mating occurs at the nest is probably because swifts need an alternative in case bad weather makes it difficult in the air. On the other hand, many of the nesting holes are so cramped that in good weather it is probably easier in flight. As already mentioned, no other birds are known to mate in the air, but few have the powerful flight of swifts or are so at home on the wing.

Aerial mating seems confined to the period shortly before the clutch is laid, but the downward glide with wings held up, and the ensuing flight with quivering wings, are not so restricted and can be seen throughout the summer, usually near the nesting colony. Very possibly, yearling birds are involved. Both the downward glide and the quivering flight can be given in turn by both members of a chasing pair. In another aerial display, one bird some fifty feet above the other dives steeply down to it with extremely rapid and powerful wing-beats, checking just before it would strike, and then being carried steeply upwards by its momentum. The meaning of this display is not known. The power of the swift in the air is nowhere seen to greater advantage.

Another, and the best known, of the aerial displays of the swift is the screaming party round the nesting colony. This is restricted to periods of fine weather, when it can be seen at any time of the day, but especially in the evening. As mentioned in the last chapter, when a 'banger' is flying round the tower, the owners of nests tend to return to them to defend them. In contrast, when a screaming party flies past, the owners of nests often come out and join in, particularly in the evening. The parties evidently have no aggressive intent. They also seem unconnected with courtship, since they occur throughout the breeding season, indeed they are most intense shortly before the birds leave in late summer, and play a definite part in the start of migration. At other times they perhaps serve to unify the group, but the social life of birds has been so little studied that we would not care to be more precise. While their meaning is obscure, these screaming parties provide one of the most characteristic sounds of an English village on a warm summer evening. A remarkable photograph of such a party is shown in Plate 6.

Since the point has been disputed for the common swift, it is valuable to find that coition has been seen both at the nest and in the air in two other swifts in the same genus, the alpine swift and the African white-rumped swift. In the alpine swift, a pair was sitting on a window-ledge, the female then flew off calling, and quivered her wings in a manner similar to that described for the common swift, the male followed her, alighted on her back, and coition followed. Aerial mating has also been seen in the house swift in Malaya, while mating has been seen at the nest in the African palm swift and the white-bellied (or glossy) swiftlet in Java. In North America, R. B. Fischer informs me that he has seen chimney swifts mating at the nest. Rapid chases through the air, resembling those of the common swift, have also been described in this species, Vaux and black swifts, but there appear to be no definite records of aerial mating. The latter has, however, been described in the white-throated swift, in which the pair are said to meet by coming from opposite directions, then gripping each other and turning over and over in the air, pin-wheeling slowly downwards for several seconds before separating. This is so different from what happens in other swifts that it might be doubted, but it has been reported by several different observers, one of whom twice collected the birds in the act, in both cases finding that they were a sexually mature male and female.

Chapter 5

THE NEST

In modern England, and for that matter in much of Europe, swifts nest mainly in buildings. They particularly favour holes in the stonework of high walls, as in old towers, and also the angle between the eaves and the wall of a house, by which they enter to nest among the rafters. Where roofs are still thatched, they often use the holes made in the thatch by house sparrows. Occasionally, they nest under the eaves in the mud nests of house martins; it is remarkable that they can fit into so small a place.

The nests are usually more than twenty feet above the ground, sometimes more than a hundred feet up, but some of those near Oxford were only ten feet up, and in Switzerland I have seen nests in a wall supporting a terraced walk in a public park which were effectively at ground level. Always, the birds have an open space in front for a clear flight in, and also a drop immediately below the nest which allows a rapid take-off.

The natural nesting place is a hole in a cliff. Nests have been reported from sea cliffs in Devonshire, Yorkshire, Norfolk and the rocky coasts of Wales, Scotland and Ireland. There are also records from natural inland cliffs in the Pennines at Malham and in Dovedale, and also in quarries excavated by man. The oldest record of quarry-nesting is from Gilbert White, who in 1774 recorded swifts breeding in the chalk pit at Odiham in Hampshire. They were still nesting there at the beginning of the twentieth century, though they have since left. Swifts have also nested in a sand quarry in the burrows made by sand martins.

There are two old British records of swifts nesting in holes in trees. About the year 1835, a Scottish naturalist saw swifts flying over pine trees in the ancient forest along the river Beauly in Ross-shire, and gamekeepers assured him that swifts nested there in old woodpecker holes. What remained of the old Highland Forest was later destroyed and the great spotted woodpecker became extinct in Scotland, so this habit must have died out. There is a similar gamekeeper's report, but with less corroborative evidence, from Devonshire in the nineteenth century, but there are no modern records.

The swift still nests in trees in various parts of Germany including the Spessart and Rhön, and there are old records from Bohemia, but the habit seems much less common than in the past, probably because old trees are now rare. The birds used oaks and beeches, and they have also bred in nesting boxes placed on trees, a habit not known in Britain. In Lapland, old woodpecker-holes in pines are the swift's usual nesting site, and one of our happiest memories is of watching a small colony in the open pine forest near Kiruna in Swedish Lapland, north of the Arctic Circle. Despite the discomfort of huge numbers of midges, we observed them for several hours as they twisted and screamed through the openings among the trees, just as they do outside the museum at Oxford, but their companions were Lapp tits and Siberian jays instead of parked cars and undergraduates.

In the tower, each pair has normally placed its nest at the back of the box, as far from the entrance as possible. Knowing this habit, we at the start put a small ring of straw at the back of each box, and many pairs later used this for lining the nest, while others used it for one side of the nest. The same nest is used and added to year after year, so that it gets gradually larger.

Nest-building often starts on the day that the second member of the pair returns from the south, though sometimes not until several days afterwards. Both male and female collect material and both build it into the nest. They work independently and do not help each other. Thus if one bird is on the nest when the other arrives with material, the sitter merely moves off to let the other build. On two occasions when both adults arrived together with material, there was some display and even a brief scuffle before each independently built it in.

Unlike what happens in most birds, building does not stop when the eggs are laid, but continues right through incubation. As a result, the nest is larger and neater at the end than the start of incubation. An adult relieving its mate on the eggs often enters with new material, and the incubating bird spends much time pecking round the outside of the nest and sticking down loose bits. So soon as the young hatch, building stops, but parents which lose their eggs continue building until late in the summer. Likewise the non-breeding pairs, mainly yearlings, continue building right through the summer. They spend much less time than the incubating birds in sticking material down, so that, in comparison, their nests are larger but less tidy. Although the yearlings do not lay eggs, they usually return to the same nesting hole in the following year, so that the time spent in building is not wasted.

The swift normally collects all its nesting material on the wing. Hence building occurs irregularly, being most frequent when there is enough wind to carry material up into the air (Plate 7). Dead grass, hay, straw, dead and also green leaves, flower petals, winged seeds, seed fluff, bud sheaths, cocoons, feathers and scraps of paper, including a bus ticket, have been found in the tower nests. The birds have sometimes brought fresh poppy petals, which made a vivid splash of red in the nest. One bird even brought a cabbage white butterfly, not as food, for it did not try to eat it, but as nesting material, which it tried to stick to the side of the nest. This it found hard to do as the butterfly was still partly alive and started a reflex jerking of its wings, but it eventually succeeded.

In collecting nest-material, as in feeding, swifts are great opportunists, using whatever happens to be common. We knew when a grass field near the tower was being cut, as the birds were soon bringing hay for their nests. Again, just after two pigeons had fought on the roof of the tower, a swift entered with a pigeon's feather in its bill. In the war, both in Italy and Denmark, nests were sometimes built of the shreds of tinfoil dropped by the RAF to confuse enemy radar.

The swift carries nesting material chiefly across its mouth (see Plate 7), from which feathers or straw often project out sideways, but it brings smaller objects inside the mouth. It sticks this material to the nest with saliva, which is used from the start of building. When producing saliva, it sometimes continues to hold the material in its bill, but at other times it first lays it down. It crouches with the head held low, and sometimes it nods its head or quivers the body, with wings held partly out. Bill and throat can be seen moving, and saliva appears in sticky threads. The bird usually does this for three or four minutes after bringing material into the box, with pauses of up to half a minute to rest, when it often lays its head on the side of the nest. It shapes the nest by turning round in it and scrabbling with its feet, eventually making a shallow cup.

All other swifts, like our bird, have to solve the problem of a nesting place which is safe from enemies and has a clear space in front for flying in at high speed. The means which the different species use are varied and it is hard to know which to admire most.

The close relatives of the European species, in the same genus *Apus*, breed mainly in holes in rocks or cliffs, while several of them, like our bird, have secondarily taken to buildings. For instance I have seen alpine swifts nesting both on a steep cliff at 12,000 feet on the wild slopes of Mount Kilimanjaro, and also in a church in the centre of a Swiss town. In May 1954, I was taken by Hans Arn, town architect, under the roof of the wonderful Jesuit Church in Solothurn. Below was a baroque triumph in white and blue with elaborate hand-carved plasterwork, while around us in the dusty woodwork above the curved ceiling were a hundred and fifty pairs of one of the most magnificent fliers in the world. From below came Bach's music on the organ, but it was almost drowned by the clamorous trilling of the birds, which conveyed the same excitement as a sea-bird colony.

The nests were on the floor or attached like brackets to side-walls. Whereas in the common swift each pair usually has its own entrance from the outside, here several pairs entered by the same gap, and one bird had often to crawl close past a neighbour, at which the trilling would be redoubled. Some of the nests were three feet down from the side-walls and to reach them their owners, after entering, unexpectedly turned sideways and let go, sliding this distance down the slope. The alpine swift is larger than the common swift, but for the most part it brings much smaller material for its nest, chiefly the outer cases of tree-buds. These are brought packed in the back of the throat, a method which the common swift uses less often, and they are then worked with saliva into a cement-like material. Feathers are also brought, held crosswise in the beak.

Five other species of *Apus* have been found nesting in rock crevices, namely the pallid swift of the Mediterranean region, the black, the mottled and the white-rumped swifts of southern Africa and the Pacific white-rumped swift of Asia. Like our own bird, they make a simple cup from plant fragments and feathers, glued together with saliva. Two other African species behave rather differently, the Horus swift nesting in sandbanks in the holes made by sand martins or starlings, and the little house swift building a small bag-shaped nest with a short tubular entrance, placing it under an overhanging rock, or on the underside of the roof of a cave, where adjoining nests often touch each other. Four of these species, the pallid swift, the African and Pacific white-rumped swifts and the house swift, at times take over the nests of swallows or martins, and all four also nest in buildings. Here the situation corresponds with the natural one, those that nest in rock crevices choosing small holes, while the house swift places its nest under the eaves of a house or under the domed roof of a mosque.

Another big group of swifts, found in America, Africa and Asia, comprises the spine-tails (*Chaetura.*) As shown in Fig. 3, these nest very differently, fixing a bracket-shaped nest of small twigs, glued together with saliva, on to the inside wall of a hollow tree. The birds usually enter by diving down from above and leave by flying up the inside of the tree to the opening. Several of the American and one of the African species have taken to nesting down chimneys, which are the nearest artificial equivalent to their natural choice. The North American chimney swift, as its name suggests, now nests far more

Fig 3: Chimney swift (*Chaetura pelagica*) at nest

commonly down chimneys than in trees, so it is likely that, as in our own bird, the change from primeval forest to cultivated land with towns has caused an increase in its numbers.

Like most other swifts the spine-tails collect their nesting material on the wing, but since twigs are not normally airborne, more drastic methods are needed. The chimney swift dives powerfully towards the top branches of a tree and snaps off small twigs with its toes as it passes, a remarkable feat. The largest of the spine-tails, the giant swift of Malaysia, differs in nesting from the others since, though it descends into huge hollow trees, it places its nest in the debris at the bottom of the hole.

Another group are the cave swiftlets (*Collocalia*), found in South-east Asia and the islands of the Pacific and Indian Oceans. Most of them nest in big colonies in dark caves, some by the shore and others far inland. In the Andaman Islands one species nests in sea-caves which are covered to the roof at each successive wave, the

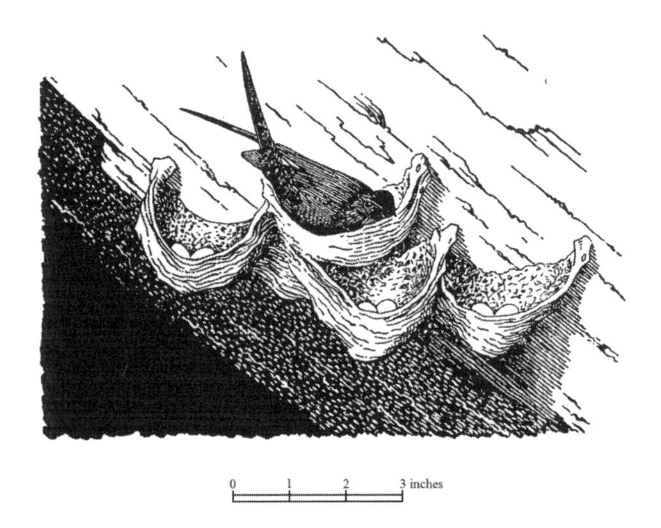

Fig 4: Nest of cave swiftlets (*Collocalia*)

birds hanging around outside to dash in as the water recedes. Quite as remarkable is a Himalayan form, found nesting in limestone potholes and underground caves at 9,000 feet above sea level. In one pothole 243 feet deep, the nests started 50 feet from the top and continued to the bottom, and the swiftlets entered by spiralling downwards in near-darkness. Others were in a cavern reached from a pothole 110 feet deep and only 15 feet across. It is amazing that the birds do not hurt themselves in collisions.

Some of the colonies are huge. Thus shortly before the birds entered for the night, an observer at the Niah caves in Borneo estimated that there were about a million and a quarter swiftlets milling around like black snowflakes. At the same time, huge numbers of bats were leaving the caves to feed, and attracted by this double feast were various birds of prey, taking their fill before dark. One species of swiftlet, the white-bellied or glossy, differs from the rest in breeding in small groups, not large colonies, and though it at times nests in caves, it often chooses more open rocky sites, even large hollow trees, and also buildings. One other species, the grey-rumped swiftlet, also nests in houses at times. A small part of a colony is shown in Fig. 4.

Swiftlets, like spine-tails, build a bracket-shaped nest on a vertical wall, but the material is different. Saliva is used not merely to stick down the materials but as a major part of the nest, indeed in some species it is the only constituent. The nest is then nearly white and looks rather like water-glass, and it is one of the few birds' nests of commercial value, since it is the main ingredient of birds' nest soup, prized by the Chinese. While the 'white nests' are the most valuable, 'black nests' are also sold. These are made by different species, which add moss, grass or feathers to the saliva. Like other swifts, they are opportunists in what they collect, and in the Andaman Islands one species used once a month to bring in human hair, after the convicts' regulation haircut.

The edible nests were brought to Europe in the seventeenth century by the Jesuits and there was much speculation as to their composition. John Ray, for instance, in the first bird book written in English, said: 'In the Sea-coast of the Kingdom of China, a sort of small parti-coloured bird of the shape of Swallow, from the foam or froth of the Sea-water dashing against the Rocks gather a certain clammy glutinous matter, perchance the Sperm of Whales or other fishes, of which they build their Nests. These nests are esteemed by gluttons great delicacies, who dissolving them in Chicken or Mutton broth, prefer them far before Oysters, Mushrooms or other dainty morsels.' In fact the material is pure saliva, and a Chinese biochemist has now shown that it has little nutritive value.

In Borneo, the nesting caves are jealously guarded and though at one time there was much surreptitious pillaging, the industry is now highly organised, to the advantage both of the natives and of the birds, which have much increased. In the nineteen-forties, each nest might be sold to a Chinese trader for 6d. and a native might make £1,000 in a season. The number of nests taken can be gauged from the fact that, at the Baram caves alone, 1,136 lb. were removed in one year, there being about 36 nests to 1 lb. At the Niah caves, already mentioned, the cathedral-like entrance-hall, several hundred feet high, is covered with ladders that give access to the remoter passages where many of the birds nest. Further inside, the natives climb by flimsy stick ladders to precarious platforms and with long bamboo poles, which carry a beeswax candle near the iron tip, they knock down the nests for an assistant to collect from the ground. To quote from E. Banks, from whom this account has been taken, 'One's impressions on entering the caves are first the soft mass of guano some ten feet thick under foot, over which run countless cockroaches, then the apparently perilous position of the climber suspended in semi-darkness so little altered by his firefly-like light, and lastly the rising and falling twittering of the disturbed swiftlets.'

So valuable are the nests that where the grey-rumped swiftlet has taken to nesting in the rooms of houses in Java, the owners cheerfully abandon the rooms and even the whole house to the birds, and soon collect enough money for a new house and much more besides. The birds can be studied more easily in houses than caves, and for this reason one observer in Java was able to discover that the white nests of the grey-rumped swiftlet were built entirely at night, and that the birds took from four and a half to six weeks to complete them. But the white-bellied species, which mixes plant material with saliva, builds by day and takes only about three weeks.

Some of the cave swiftlets nest where it is almost, if not completely, dark. This raises the question of how they find their way in and out without hitting the rocks or each other. The constant 'twittering', mentioned by all visitors to the caves, probably provides a clue. It has been proved that bats emit supersonic squeaks and detect the echoes of them given off by solid objects, and that this is how they avoid obstacles and find their prey. Similarly the oil bird or guacharo, a South American relative of the nightjars, roosts by day in completely dark caves, coming out to feed at night, and it finds its way through the caves by the echo-location of a sharp clicking note. This click, unlike the squeak of a bat, is audible to the human ear, having a frequency of about 7,000 cycles per second. The individual click lasts for about a thousandth of a second and the bird utters the clicks in short bursts, with pauses between. Oil birds released in a completely dark room flew freely without hitting obstacles, but when their ears were blocked up so that they could no longer hear the echoes of their clicks, they at once flew into the walls. It seems possible that cave swiftlets have a similar mechanism. The principle, of course, is that familiar to us in the tapping of a blind man's stick.

The white-throated swift (*Aeronautes*), a western American species shown in Fig. 5, nests in crevices in cliffs, with an apparent preference for precipices where the rocks are liable to give way, so that few men have climbed to them. The nest is a simple cup of feathers. Occasionally, this species has nested on a building or in the nest of a martin.

Fig 5: White-throated swift (*Aeronautes saxatilis*)

The black swifts (*Cypseloides*) of central and western America attach a nest of mud and moss, lined with fern-tips, to a rock wall. Usually, though not always, the nest is placed above or close to water, and sometimes the birds actually fly through a waterfall to reach their nests. One species, the black swift of western North America, at times nests on steep sea cliffs, and as it lays only one egg, its discoverer was at first thought to have mistaken the nest for that of a petrel. Later, the same species was found nesting behind waterfalls in the mountains. The collared or cloud swift of tropical America,

Fig 6: Collared or cloud swift (*Cypseloides zonaris*)
nesting behind waterfall

shown in Fig. 6, also flies through a curtain of falling water to its nest, and the curious may note that the first flight that the young swift ever takes must likewise be through the water, though this has never been described. One British bird, the dipper, also nests behind waterfalls, but as it takes its food from the water, this is less surprising than for a swift to do so. Most swifts, as already mentioned, collect their nesting material on the wing, but it may be doubted whether black swifts could collect mud and moss in this way. The birds have been seen alighting behind waterfalls where no nests were present, and they perhaps collect their nesting material clinging to rocks, meanwhile protected from possible enemies by the falling water.

The American palm swifts (*Tachornis*) nest in the hollow tube formed by the long, hanging leaves or by the flower spathe of a palm tree. One species and its nest is shown in Figs. 7A and 7B. The birds enter by flying up from below, and the nest, which has its entrance

Fig 7A: New World palm swift (*Tachornis squamata*)

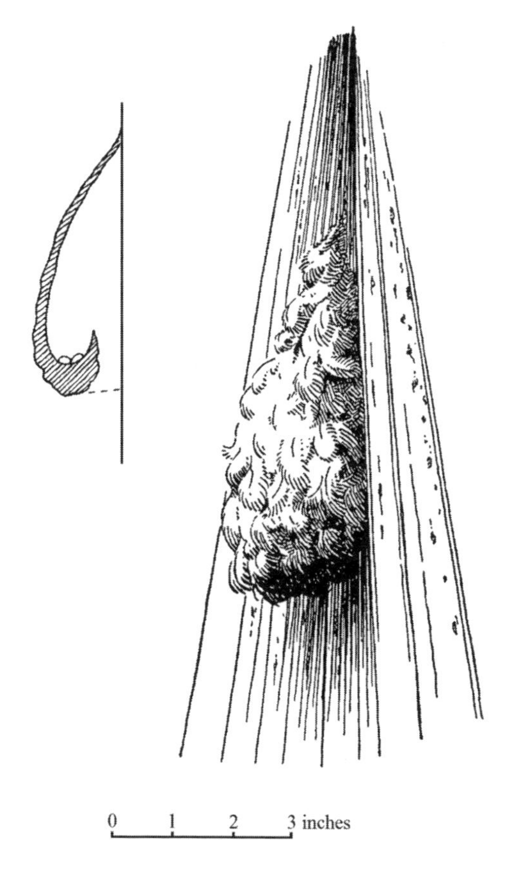

Fig 7B: Nest of New World palm swift

at the bottom, is a small bag of plant fibres and feathers, which may be worked with saliva into a felt-like material. This nest was first described in Jamaica over a century ago by Philip Gosse, who is now chiefly known for the entry in his diary: 'E. delivered of a son. Received green swallow from Jamaica.' Actually he was a fine naturalist, though his fundamentalism involved him on the wrong side in the argument over Darwinism.

The most elaborate nests built by swifts are those of the beautiful scissor-tailed swifts (*Panyptila*) of tropical America, shown in Fig. 8B. The nest is hung from the underside of a branch or overhanging rock and consists of a tube of from seven inches to over two feet in length, with the entrance at the bottom and a wider chamber at the top, in which there is a side pocket or shelf which holds the eggs. The

Fig 8A: Scissor-tailed swift (*Panyptila*)

nest is made of dried feathery tufts of plant seeds or birds' feathers and is worked with saliva into a close felt. It is so well made that it may be used for several years in succession, the birds merely adding a new shelf inside. In plan and construction the nest is similar to that of the American palm swifts except that the birds add their own long entrance tube. Similar long entrance tubes are found in the nests of various other tropical birds, notably some of the weavers, and they are thought to be a device to prevent animals from entering to rob the eggs. One species of scissor-tailed swift, like so many other swifts, has taken to nesting on buildings, so that these curious structures can be seen hanging under archways in Trinidad and from ceilings in Panama.

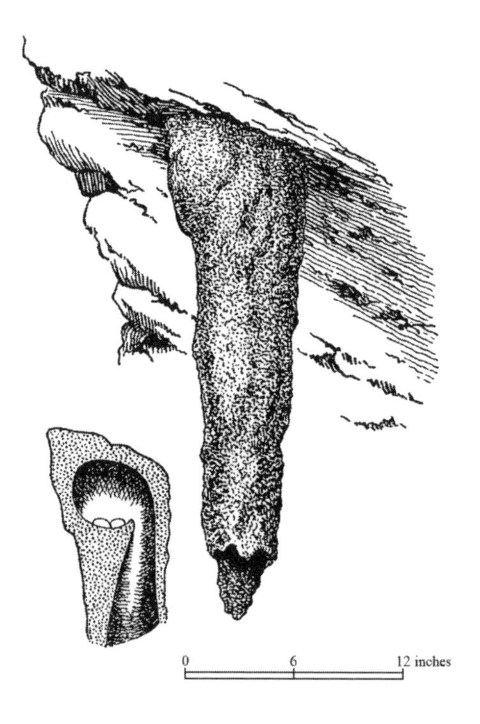

Fig 8B: Nest of scissor-tailed swift

The Old World palm swifts (*Cypsiurus*) nest very differently from the New World forms. The nest is a simple strip of feathers or plant fibres, in a shape like that of a shallow bowl of a spoon with the longer axis vertical, about 1½ inches across and with a small projecting rim at the bottom. This is fixed to the vertical side of a hanging palm leaf

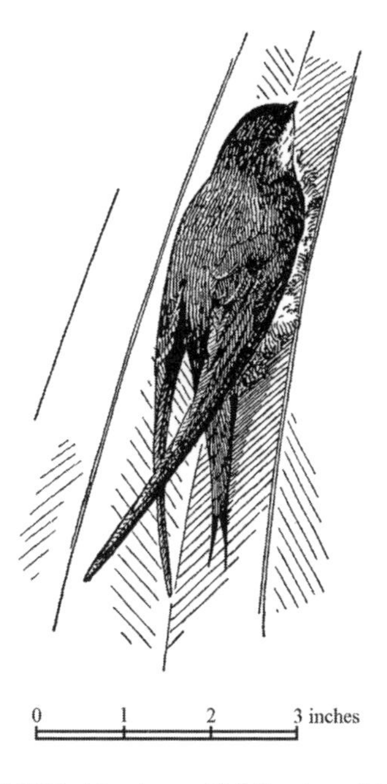

Fig 9A: Old World palm swift (*Cypsiurus*) incubating
(redrawn from W. M. Moreau)

on the inner (technically the under) side, the eggs being placed on the small projecting rim. 'When the wind blows the cradle will rock', and to prevent the eggs from falling, they are glued to the nest with saliva. An African observer trained by R. E. Moreau in Tanganyika actually saw the bird stick the egg down, vomiting saliva and putting it alongside the egg, all the time pressing firmly with its body to hold the egg in place. As is well known, most birds regularly turn their eggs during incubation, though the reason for this is not properly understood. These palm swifts cannot turn their eggs, yet they hatch safely. They also differ from most other birds in their attitude when brooding the eggs, not sitting horizontally across the nest, but clinging vertically, as shown in Fig. 9A. Incidentally in Asia they nest regularly in the roofs of native houses thatched with palm.

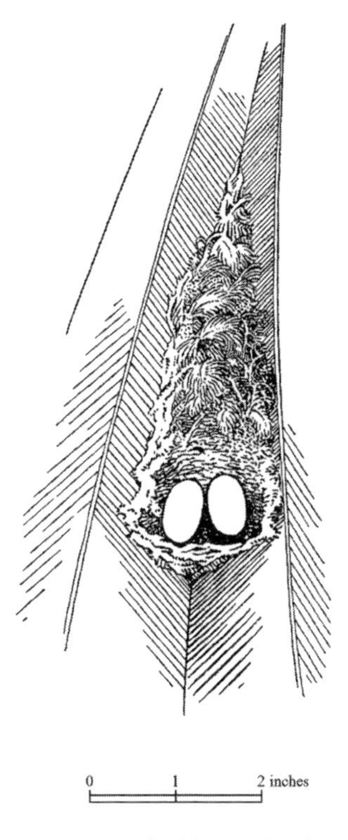

Fig 9B: Nest of Old World palm swift

While the scissor-tailed swifts of tropical America build the most elaborate nests, the prize for ingenious simplicity must be given to the crested swifts (*Hemiprocne*) of South-east Asia.

Though they are large, one species being much bigger than our own swift, the nest is little more than an inch across. It is literally an egg-cup, being just large enough to hold the single egg. It is placed on the side of a small branch, usually about forty feet from the ground, and is lightly made of feathers and strips of bark. So frail a structure could not support the bird, but the top of the nest is flush with the top of the branch, and the bird lets the branch take its weight while keeping its brood-patch in contact with the egg, as shown in Fig. 10. The egg, like that of the Old World palm swift, is glued down with saliva.

Fig 10: Crested swift (*Hemiprocne*) at nest

This survey shows the diversity of adaptation in the nesting of swifts, our own species in holes in cliffs, the spine-tails diving into hollow trees and breaking off twigs in flight, the swiftlets seeking dark caves and making the whole nest of saliva, black swifts dashing through waterfalls, American palm swifts flying up into palm spathes, scissor-tailed swifts building their own two-foot tube of felted feathers, Old World palm swifts gluing the eggs to a small shelf of feathers and clinging vertically to brood them, and crested swifts making a tiny egg-cup on the side of a branch across which they sit.

Perhaps because we were told the same instances too often, we of this generation have tended to cease marvelling at the wonders of nature; but, as this fresh example may show, the marvel is always there. It is not surprising that our orthodox great-grandparents followed Paley in believing that the existence of such intricate designs proved the existence of a Designer, while our less orthodox grandparents, only partly persuaded by Darwin and Huxley, felt that natural selection could not be enough, and found relief in the idea of a Life Force. But that dispute will be considered later.

APPENDIX

SUMMARY REVIEW OF THE WORLD'S SWIFTS

The following summary classification may prove helpful to readers of the preceding and other chapters. Except for three species of crested swifts (*Hemiprocne*), all swifts look very similar and are placed in one family, the Apodidae. The latter is divided into the Apodinae, which have the hind toe pointing forwards, and the Chaeturinae, which have the hind toe in the normal position pointing backwards. In the following list, the arrangement of the genera is partly original, for reasons published elsewhere.

APODINAE

1. *Apus*: Typical swifts, comprising 10 Old World species, 8 of which breed in Africa, 5 in Asia and 3 in Europe (namely the common swift *A. apus*, alpine *A. melba* and pallid *A. pallidus*). Most species are sooty black, some brown, nearly all have pale throats, some have white rumps and one mainly white underparts. Tail moderately forked, average wing-length varying from 12 to 23 cm, most nest in holes in cliffs or buildings, usual clutch 2 or 3.
2. *Cypsiurus*: Old World palm swift, 1 species in tropical Africa and Asia, sooty with pale throat, delicately forked tail, wing 13 cm, open nest on palm leaf, clutch 2 or 3.
3. *Aeronautes*: White-throated swift (western North America) and two South American species, black with patches of white on underparts or neck, slightly forked tail, wing 12 to 14 cm, nest in holes in cliff, clutch (one species) 4 to 6.
4. *Panyptila:* Scissor-tailed swifts, 2 species in tropical America, glossy black with white throat, neck and edge of wing, delicately forked tail, wing 12 and 19 cm, nest a hanging sleeve under branch or rock, clutch probably 3.
5. *Tachornis*: New World palm swifts, 3 species in tropical America, sooty with pale underparts, well-forked tail, wing 9 to 10 cm, nest a bag in palm foliage, clutch 3.

CHAETURINAE

6. *Chaetura*: Spine-tailed swifts, some 17 species in Asia, Africa and America, including chimney swift (*C. pelagica*) of eastern North America (which is about half the weight of common swift), and the large needle-tailed swift (*C. caudacuta*) of north-east Asia, which has wandered to Britain. Most are sooty with pale throat,

a few are dark brown, or glossy black, blue or green, some have white on rump, tail or abdomen. Tail square-ended with prominent spiny tips to feathers, wing varying from 11 to 20 cm, nest inside hollow tree or chimney, usual clutch 3 to 5.

7. *Collocalia*: Cave swiftlets, some 20 species in south-east Asia and islands of Pacific and Indian Oceans, most sooty all over, some glossy, some with white on rump or abdomen, tail square-ended or slightly forked, wing 9½ to 16 cm, nest largely or wholly of saliva, in caves, clutch 2.

8. *Cypseloides*: Black swifts, 9 species in tropical America, but one (*C. niger*) extending up west coast to Alaska, most black all over, but some have white and one a red collar, tail square-ended or slightly forked, in most with slight spiny tips, wing 12½ to 23 cm, nest on cliffs near water, usual clutch 1.

HEMIPROCNIDAE

9. *Hemiprocne*: Crested swifts, 3 species in south-east Asia and islands of western Pacific, perch freely on trees, wings being less specialised than in Apodidae. Wings metallic blue, body grey with some white, white stripes and chestnut patch on head, small crest of feathers, tail delicately forked, wing 13 to 22 cm, nest a small cup on side of branch, clutch 1.

Notes: (i) This classification, in part my own, is revised and much simplified from that in *Peters' Check-List of Birds of the World* (1940).

(ii) There are occasional exceptions to some of the summarised statements about each genus.

(iii) Wing-lengths refer to the standard measurement from the carpal angle to the tip of the longest primary. The statement 'wing 12 to 23 cm' means that the smallest form in the genus has an average wing-length of about 12 cm and the largest of about 23 cm.

Chapter 6

LAYING AND INCUBATION

In the tower, the eggs are laid about a fortnight after the second member of the pair has returned, commonly in the last week of May but sometimes in the previous week and sometimes in the first week of June. Regular visits during the laying period showed that nearly all the eggs were laid in the morning, after 8 a.m. and before 11 a.m., though a few were laid earlier and a very few in the afternoon. On the morning when she will lay, the hen perhaps does not go out beforehand, as this is the only time when we have found the droppings of an adult bird in the box.

We have several times watched the hen bird actually laying, though as she sits closely, there is not much to see. On one occasion the cock was present at the start and kept trying to sit on the first egg, which was already laid, but the hen would not allow this. At 10.15 a.m. the cock left. The hen sat quietly until 10.43, when she humped her back and put down her head as if looking under her wings, which were held low. She then resumed her normal sitting position, and when she preened at 11 a.m. we could see the second egg under her. At 11.04 she turned the eggs with her beak and seven minutes later went slowly to the entrance and flew out. The humped attitude was also seen in other instances, and perhaps indicates the moment when the egg is actually laid.

The clutch consists of two or three white eggs. The eggs are white in most other birds which nest in dark places, such as sand martins, kingfishers, woodpeckers and owls, and this is an obvious adaptation to help the parents to see them. It is interesting that crested swifts, which do not nest in the dark, differ from all other swifts in that the egg is not white but pale blue.

The egg of the swift weighs about 3½ grams, one-twelfth of the weight of the bird. When food is scarce, its production may well impose a strain on the hen bird, and this is borne out by the marked influence of the weather on the start of laying, also on the interval between the laying of the eggs of a clutch and on the number of eggs laid.

In most years at the tower the first egg was laid between 17 and 22 May. But in 1951 the first half of May was cold with fairly strong north-easterly winds, and the first warm and sunny day did not come until 21 May. The first egg was laid on 26 May, unusually late and five days after the first warm day. In 1950, likewise, the middle of May was cool and sunless, but 18 May was warmer and sunny, and the first egg was laid in the tower five days afterwards. The same happened again in 1955, when the first egg was laid as late as 28 May.

The influence of the weather was shown more strikingly in 1949, when many pairs had started their clutches between 22 and 28 May, but there was then a gap of eight days and the remaining pairs did not start until 5 June. The weather records showed that between 24 and 31 May there were unusually strong south-westerly winds, which reached force 5 on the Beaufort scale on one day and force 4 on several days. The wind did not drop until 1 June. In strong winds there are few air-borne insects, and it will be noticed that the gap in laying started five days after the first day of strong winds, and ended on the fifth day after the wind dropped. Much the same happened in 1950, but owing on this occasion to a sudden fall in temperature. This occurred on 24 May and the next warm and sunny day was 28 May. Several pairs had started their clutches between 23 and 28 May, but no others until 2 June. Again, therefore, laying stopped five days after the start of bad weather and was resumed five days after the start of good weather.

The story now seems simple, but we must confess that we did not appreciate the cause of these happenings until we read of a

Dutch worker's discovery that the start of laying in the great tit is influenced by the weather four days previously. Examination of our swift records then showed a similar effect, except that the interval was five instead of four days. The reason is evidently that it takes a great tit four days and a swift five days to make a complete egg, and once this process has started it is not stopped.

If a swift has already started its clutch, it usually lays the second egg whatever the weather. (Otherwise the first egg would be wasted.) But bad weather influences the interval between the eggs. Most song-birds lay one egg each day until the clutch is complete, but in the swift the usual interval is two days. This in itself suggests that the bird may find it hard to collect enough food to make the eggs. Moreover in a proportion of the clutches the interval between the eggs was three days, and analysis showed that a three-day interval was twice as common in cold as in warm weather.

The weather also influences the number of eggs in the clutch. In the first week of the laying season (18 to 24 May) most clutches consist of three eggs, but the proportion varies with the preceding weather. If it has been warm, nearly all the clutches consist of three eggs, whereas if it has been cold only about half of them do so. Likewise in the last week of May, when the average clutch is smaller, if the preceding weather has been warm about half the clutches consist of three eggs but if it has been cold scarcely any do so. The basic reason why the swift lays a clutch of two or three eggs will be considered later, in Chapter 16, and the only point made here is that the weather just before laying has a subsidiary influence, presumably because when food is short the production of a further egg might impose too much strain on the bird.

Incubation by day usually starts when the clutch is complete, though in clutches of three eggs sometimes with the second egg. The sitting bird normally has its back to the entrance (Plate 1). It is continually fidgeting, preening, scratching, quivering the body, shuffling the eggs with its feet or turning them with its bill, rattling its bill on the side of the box, picking up any loose nesting material within reach or building it into its nest. If a 'banger' is heard outside, the sitter may temporarily turn towards the entrance, and sometimes it even leaves the eggs and looks out of the hole (Plate 2), but it usually returns to the eggs after a few minutes. One bird stood up over its eggs each time that it heard the loud clapping of the crowd in the near-by Parks as successive New Zealand wickets fell to a triumphant University (which was the only team to beat the tourists in that year). In very hot weather, the parent sometimes sits beside the eggs instead of over them.

The two parents take turn and turn about on the eggs. The sitter often greets the arrival of its mate with a scream or with mild threat display as it moves slowly off the nest. Sometimes it is reluctant to leave and the newcomer then prods it gently or gradually insinuates its body under that of the sitter. On one occasion when the newcomer was kept waiting, it picked up some grass lying in the box. This is an instance of a 'displacement activity', a type of action meaningless in itself but which Dr Tinbergen has shown to be a regular feature of the behaviour of birds, and for that matter of human beings, when frustrated in what they urgently want to do. At night, and in heavy rain by day, both parents sit in the box, but we do not know if they periodically change places on the eggs under these circumstances.

In 1949, we on four days watched continuously in the tower from dawn till dusk, relieving each other in turn so that we could get meals – at about the same average interval and for the same purpose as the swifts that we were watching. It is, naturally, dull to watch at a nest where no change may occur for several hours, but in the upper part of the tower we could record at a dozen nests at the same time. Not all were in sight from one place, but the returning parent makes a slight bang as it enters, and on hearing any noise we checked at the box in question. In the two following years, we watched only from 8 a.m. to 6 p.m., which proved much less tiring and was nearly as rewarding.

The parent birds change places about once every two hours. The sitting bird rarely leaves before its mate returns, and the interval evidently depends mainly on the time needed by the bird that is out hunting to collect enough food for itself. Thus the interval is usually longer before 11 a.m., when the feeding bird is getting its first good meal of the day and when insects are scarce, than it is later in the day, when insects are more numerous and when the bird has already fed several times. The longest interval for which one bird sat without relief was six hours, the shortest was only two minutes. In the latter case, the bird which had been relieved returned at once, so one may suppose that it was not specially hungry and still had a strong urge to brood. The highest number of reliefs observed at one nest in a ten-hour watch was twelve and the lowest two. We found that some pairs regularly relieved each other after a shorter interval than others, perhaps because they were quicker at catching food than others, since the same individuals later tended to bring food somewhat more frequently to their young. We also found that nest relief occurred at more frequent intervals on some days than others, but were unable to account for this variation. Contrary to what has sometimes been stated, we found that both sexes took an equal share in incubation, and this was also the experience of the Swiss observer Weitnauer.

The most interesting point that came out of these long watches was that the sitting bird sometimes left its eggs and went out to feed before its mate came back to relieve it. The eggs were then left uncovered for a period which varied from a few minutes to six and a half hours. This was liable to happen in cold or windy weather, and was presumably due to the sitting bird getting hungry before its mate returned. The eggs were left uncovered in this way chiefly between 11 a.m. and 6 p.m., which is the period when air-borne insects are most numerous and the parent birds can feed most easily. Various African swifts have also been found to leave their eggs uncovered for long periods of the day, but this is less surprising as they experience much warmer weather than in England.

It is popularly believed that if a bird's egg is left to get cool during incubation the embryo will die, and that the period just before the egg hatches is particularly critical. To our surprise, those swifts' eggs that were left to cool hatched normally. This applied even to two eggs which were left uncovered for six hours on the day before they hatched. It should be stressed that the eggs are left in this way chiefly during cold weather, because that is when the parents are short of food. Evidently the embryo swift in the egg is resistant to cooling, and this may be regarded as an adaptation which helps the swift to breed successfully in a cool climate. A similar ability is found in the naked nestling swift, which can likewise survive long periods of cooling in bad weather, when both parents leave it in order to collect food. We have not made experiments to see how long a swift's eggs can stay cold without harm, but one clutch which was deserted by the parents and left cold for two days was put in an incubator and later hatched successfully. It is interesting that the eggs of the Manx shearwater are also resistant to cooling. Like the swift, the Manx shearwater has difficulty in finding enough food in stormy weather and may leave the egg unbrooded. To complete this story one needs to know how long the eggs of other and more ordinary birds can withstand cooling, but I have found no precise figures on this point.

Occasionally a swift throws an egg out of its nest (Plate 12) and more rarely it throws out the whole clutch. A cracked or chipped egg is nearly always thrown out, and when we have replaced it the bird has often thrown it out again later. But many of the ejected eggs are sound, and when we have replaced them they have hatched successfully. One pair threw out one of their two eggs seven times in the course of incubation, but we replaced it each time and eventually it hatched. The Swiss worker Wietnauer has found that eggs are thrown out mainly in spells of bad weather, but this has not been our experience at Oxford, and the reasons for this curious habit need further study.

Once an egg was ejected while we were watching. It popped out from under the sitting bird's flank so quickly that we could not see how it was moved, but probably it was with the foot, as later we saw a hatched eggshell pushed from the nest in this way. We have also seen a swift pick up a cracked egg in its mouth and carry it to the nest entrance, but it then fell on the floor of the box and rolled back, though it vanished later in the day. The African white-rumped swift has been seen to carry an egg in its mouth to the nest entrance and to drop it out.

If the clutch is lost, the female may not lay again that year, but sometimes it lays a second clutch. This may follow closely after the first if it was deserted before the start of incubation, but the interval is usually two or three weeks if incubation has already begun. One exceptional pair laid five eggs in seven days, throwing out the first three but hatching out the last two.

While incubation does not usually start by day until the clutch is complete, one of the parents covers the eggs at night from the time they are laid. As a result, the first egg usually hatches a day before the second, and the second a day before the third. The time required for incubation, measured from the laying to the hatching of the last egg, is on average 19½ days. Other species of swifts incubate for a similar length of time. The period is nearly a week longer than in song-birds of comparable size.

In late May 1948 and late May 1953, the interval between laying and hatching was four to five days longer than usual. In 1948 strong winds and in 1953 sudden cold came just when the clutches were completed, presumably causing the swifts to leave their eggs unbrooded for so long that proper incubation did not begin until the start of warmer weather.

Of all the eggs laid in the tower, about three-quarters have hatched. But of those laid just before the period of strong winds in late May 1953 only one-third hatched, and of those laid just before another period of strong winds in May 1954 only about one-half hatched. These records suggest that, occasionally, incubation may be so much interrupted by bad weather that the embryos suffer. If so, since bad weather later in incubation has not had such an effect, it seems likely that the swift's eggs are most sensitive to cooling at the start of incubation.

The effects of cooling on the duration of incubation and on the addling of eggs are problems best solved with an incubator. It is the privilege of the naturalist to suggest experiments for the zoologist, he cannot himself follow up every such question that presents itself, and we gladly leave this particular problem for others, since we prefer not to risk damage to our swifts' eggs but to see them hatch, and so to study them in the later phases of their life.

Chapter 7

THE NAKED NESTLING

A young swift, like a young song-bird, is hideous. When about a week old it has a relatively large mouth and head, very short legs with large toes, and the naked greyish-pink body chiefly consists of a swollen abdomen, through the skin of which can be seen the gut and purple liver, both of great size for so small a creature. Indeed, they are so large that the cloaca, instead of being at the rear end of the animal, as it will be in the adult swift, is pushed up on to the top of the back. When well-nourished, the nestling also has patches of fat under the skin.

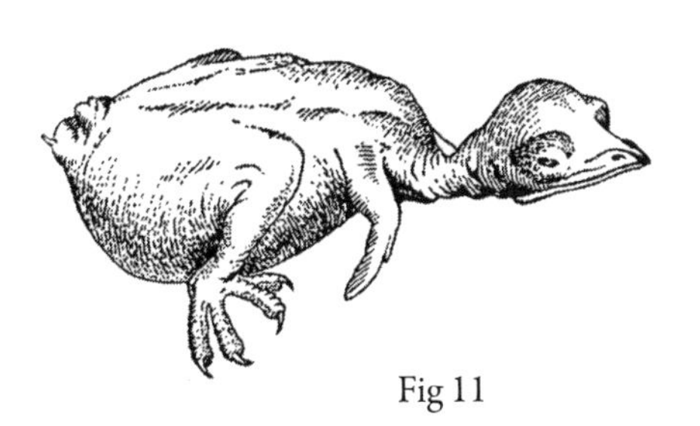

Fig 11

This distorted monster is, when looked at properly, a marvel of adaptation. The nestling lives protected in its nest-hole from the outside world, and it has one main job, to grow as quickly as possible. The large mouth helps it to take food rapidly from its parents (Plate 9), the huge gut and liver speedily convert the masses of small insects into flesh, the fat under the skin is a valuable reserve for times when food is short, and the strong feet prevent it from falling, which is specially important when the nest is on a narrow ledge with a big drop below. The toes, incidentally, are in opposed pairs, not all pointing forwards as in the adult.

At one time the form of a young animal was explained by supposing that it passed through ancestral stages in the evolution of its species. But no ancestor of birds looked anything like a young swift. The old idea contains a partial truth, but it hides a far more important one, that each young animal has to survive in its own world and that it has special adaptations to this end, many of which disappear before it becomes adult.

In contrast to the nestling swift, one of the most charming sights in nature is a brood of newly hatched ducklings or game-bird chicks, with their fluffy down, excited calls and quick movements. For this attractive spectacle we must ultimately thank the foxes and hawks. Ducklings and chicks are not protected in a nest but search for their food in the open where there are many enemies. As a result, they have evolved down which is coloured to match their surroundings, long legs and toes for quick movement, a well-developed eye and good muscular control. The development of the wings is of special interest. It is of great value for a chick to be able to fly from ground-living enemies. A young blackcock can fly when it is only ten days old, though its weight is only one-twentieth of that of the adult and it is still clothed in down except for the all-important wings. Further, whereas in most young birds the outer (primary) and inner (secondary) quills develop together, in the game-bird chick only the secondaries grow at the start, and these by themselves provide a large enough surface to lift the tiny body. The duckling on the other hand lives on the water and escapes by swimming and diving. Its wings, instead of being the first, are the last part of the body to grow feathers, and it cannot fly until it is almost two months old, hence the term 'flapper'. This marked difference between chick and duckling is readily explicable through adaptation to their different ways of life, but would be impossible to explain in terms of what their ancestors looked like.

To show the difference between a game-bird chick and a nestling swift, it is best to compare the smallest European game-bird (the quail) with the largest European swift (the alpine swift), as their young are of similar weight at hatching. In a newly hatched quail the gut makes up just under 10 per cent, the brain just over 6 per cent, of the weight of the body. In a newly hatched alpine swift, on the other hand, the gut makes up just under 15 per cent and the brain just over 3 per cent of the body. This means that the gut is 1½ times the size, but the brain only half the size, of that of a quail chick. The nestling alpine swift is primarily adapted for eating and growing, hence its large gut, but it at first has no need of accurate eyes or other senses, or of good muscular control, hence it needs far less brain.

The nestling swift is as efficient in its own world as the game-bird chick in its very different one, and it is only our human standards of beauty and ugliness which deceive us into thinking otherwise. Similar contrasts occur in mammals. Thus a lamb, like a game-bird chick, is active from birth, it lives in the open, so has a warm covering to its body, and it escapes from enemies by running, so has disproportionately long legs. On the other hand a human baby, like a nestling swift, lives in a protected place and is chiefly concerned with eating and growing. It has scarcely any hair, a relatively large mouth, sucking lips and tongue, and a loud call when hungry, while when it is well nourished it stores much fat under the skin for possible use in emergencies. Further, the hands are strong for grasping, the legs are disproportionately short and weak, and the general muscular control is much poorer than in the lamb. These features are obvious adaptations to its way of life and strongly recall parallel features in the nestling swift. But mothers will continue to think young swifts hideous, their own babies adorable, for such is mother love (a preferable term, surely, to the modern 'maternal instinct').

So soon as the first nestling of the brood has hatched, the parent swifts start to collect food for it, but the change in their behaviour from brooding to hunting is gradual. Our ten-hour watches in the tower showed that, when the young are under a week old, the parents brood them for almost the whole day, while in the second week of their life they are brooded for about half the day. After this they are rarely brooded by day, though the parents cover them at night until they are much older. An adult can bring in a large bolus of food stored in the throat (Plate 8). Its mate would normally leave and the newcomer would then take its place on the young and give them the food, remaining to brood them until its mate returned. When once this was disrupted by a 'banger' outside, the parent which had just come in turned and dashed for the nest entrance, while the other screamed from the nest.

Newly hatched swifts are normally brooded continuously, but in cold, windy or wet weather when air-borne insects are hard to find, it is necessary for both parents to be out hunting. The nestlings, like the eggs under similar circumstances, may then be left uncovered, for up to several hours. If the bad weather lasts for several days, they become cold, clammy to the touch and half-torpid. When we first saw such nestlings we supposed that they would die, but to our surprise, if food became plentiful again in time, they fully recovered. This astonishing capacity to survive cooling contrasts with what happens to a newly hatched song-bird which, despite a warmly lined nest, usually dies if left uncovered for more than an hour or two. In song-birds, as a result, one of the parents must always stay to brood the tiny young, but this does not usually matter, since song-birds usually have a dependable and abundant food supply for their young. In contrast, the food supply of the swift cannot be depended upon, but the nestlings' ability to withstand cooling enables it to tide over periods of food shortage when both parents have to be out hunting. It is more important than the similar adaptation of the embryo in the egg, since after the young have hatched, the parents need to find food not merely for themselves but also for their young, and so have to be away from the nest for longer.

The nestlings of most kinds of swifts remain naked until the feathers appear (see Plates 10, 11 and 12), but young crested and Old World palm swifts grow a covering of down a few days after hatching. These are the only swifts which nest in the open, and in both of them the down matches their surroundings, so it probably serves more for concealment from enemies than for protection from the weather. The nestling palm swift, for instance, is said to look like a clot of lichen.

As shown in Plate 8, the parent swift returning with food for its young has a large bulge below the beak, due to the mass of insects packed into the throat and stuck together with saliva. There is usually a definite ball of food, though at times the insects adhere loosely. On coming to the nest, the adult holds it head low with throat moving and then produces the food. When the nestlings are very small, the adult produces only part of the meal at a time and may divide it among the chicks. The feed often lasts for three or four minutes, as the adult sometimes takes the food back into its mouth and then produces it again, possibly because the original portion was too large. But from the time that the young are about a week old, the food is passed in one large ball to only one of them, and the feed lasts for only a few seconds. As a result, it is hard to see what happens, but photography by high-speed flash gives a better view. Plate 9 shows the parent feeding a newly hatched nestling, while Plate 17 shows an older nestling with its enormous gape. Plate 10 shows young about seven days old.

The young are normally fed on the nest itself. This remains true even when the young are older and no longer stay in the nest, but wander about the box or sit looking out of the entrance hole. At the return of a parent, these older young dash to the nest at the back of the box, and the parent also goes to the nest before giving up its food. In bad weather when the young were unusually hungry, a nestling occasionally got a meal from a parent as it alighted at the nest entrance, and the parent then immediately left to hunt for more, but such a high-speed turn-round was exceptional.

The nestling swift has a plaintive high-pitched note much feebler than the adult's scream. We heard one calling while still inside the egg, the sound being audible six feet away. Young nestlings beg for food from the incoming parent with squeaking and by waving the open bill, and when they are older they also chase the adult round the box with excited flapping of the wings, repeatedly trying to grasp its beak in theirs. The begging becomes much stronger when the young are short of food and a violent scrimmage may develop, which is continued by the unsatisfied nestling after its nest-mate has got the meal. The young usually start to beg as soon as they hear the sharp knock as the parent alights at the nest entrance, and when short of food they often react to other similar noises, such as a sudden gust of wind on the slates outside, or the distant backfire of a car in the road, or a sneeze by the observer, and then continue the reaction by begging from each other. They also tend to beg from the brooding adult when it leaves them to go out hunting (although it of course has no food to offer), and after about the first week of their life, they keep up a quiet high-pitched murmuring throughout the time that the adult stays with them between feeds, though not when both parents are away. This noise, like the begging, seems more intense when the young are short of food.

Probably because temporarily satisfied, a nestling has occasionally failed to beg for food from an incoming parent. With small nestlings, the parent then prods them gently, after which they open their mouths and are fed. With older young, we sometimes found that the parent merely waited for a few minutes, after which the nestling took the food, but in one instance when the nestling failed to respond the adult left the box still carrying the food, and on another occasion, after a pause in which it preened the nestling's throat, it swallowed the food itself and then left. We formed the impression that after such behaviour, or when the young begged very little, the parents stayed out unusually long before returning with another meal for them.

When it hatches the young swift weighs about 2¾ grams. It then puts on weight rapidly, in good weather reaching its highest weight of about 56 grams (2 oz) in the fourth week of its life. After that, if good weather continues, it falls gradually in weight while it grows its feathers. But this is the picture only in continuous fine weather, and it is greatly altered in bad weather when the parents cannot find enough food. In an English summer there are almost always several such cold or wet spells before the young swift is fully grown, and in each of these, whatever its age, it rapidly loses weight and continues to do so until the weather improves or it dies of undernourishment.

Here, then, is another adaptation of the young swift. If left without food, a nestling song-bird dies after a few hours, but a nestling swift can survive for several days or even weeks. This is partly because, as already mentioned, a young swift stores much fat under the skin when food is plentiful and uses this when it has to starve. In addition, a young swift greatly cuts down its activity when short of food, as discussed later.

Shearwaters and petrels, like swifts, cannot obtain food in rough weather, and it is interesting to find that their young, like those of the swift, can withstand several days of starvation. As a result, the weight-curve for a young petrel or shearwater looks very like that of a young swift. On the other hand, the weight-curve for a young song-bird with a dependable food supply is much steadier.

The nestling swift can starve for several days even when it first hatches. In the tower, one nestling was deserted by its parents as it was emerging from its shell, so it was never fed, and it was also left cold and unbrooded. To our surprise, it was still alive forty-eight hours later, and it may have survived for another full day, though not more. Presumably it subsisted on the nourishment remaining internally in the yolk-sac, and this must be a valuable asset for nestling swifts which happen to hatch in a spell of bad weather.

In the middle of the nestling period, when the young have large stores of fat, they can survive for much longer. As an extreme instance, the Swiss worker Weitnauer weighed twelve nestling swifts just before the start of a cold spell in 1948, when their average weight was 56 grams. In the fortnight from 18 June to 2 July, when the weather was very cold with snow or rain, they all lost so much that the average weight dropped to almost exactly half of what it had been. Probably the young would not have survived so long as this had it not been for one good day about half-way through the period, and, in continuing bad weather, most of them eventually died. More remarkably, in France a young swift taken from a nest when it weighed 57 grams and kept without food did not die for twenty-one days, by which time it weighed only 21 grams.

The occasional gains in weight of the young swift in good weather are almost as remarkable as the losses when food is short. At Oxford in early July 1948, after a long spell of bad weather, there were three warm and still days which provided excellent feeding. Several of the young put on 10 grams during the course of one day, thus adding one-fifth or even one-quarter to their weight at the start of the day.

Because of its capacity to starve in bad weather and to put on weight rapidly in good weather, the weight-curve for a young swift varies greatly with the season. Three extreme examples at Oxford are shown in Fig. 12. Bird A had fine weather throughout and grew fast and well. Bird B had fine weather at the start, when it grew almost as well as A, but later, in bad weather, it dropped heavily, losing two-fifths of its weight in all. Bird C experienced the opposite conditions, having bad weather from the second to the fifth week of its life, during which it only just managed to survive at a low level, and then excellent weather in which it put on weight rapidly. The variations in weight due to the weather are also shown by the extreme figures. Thus at ten days old, one Swiss bird which had experienced continuous fine weather weighed as much as 50 grams, whereas one Oxford bird which experienced continuous bad weather weighed only 5 grams. These variations greatly affect the rate of growth of the feathers, to be considered in the next chapter.

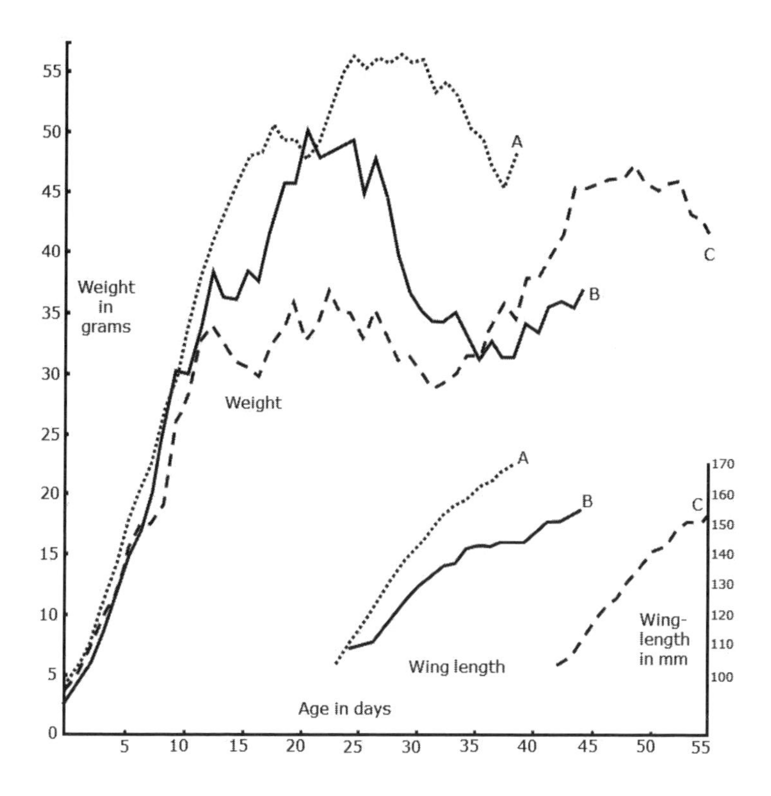

Fig 12: Increase in weight and wing-length of three nestlings, raised A in good weather, B in good weather at start, bad weather later, and C in bad weather from second to fifth week, then good weather

In Tanganyika just before the war, R. E. Moreau studied the breeding habits of the white-rumped and house swifts, which are in the same genus as our own bird. To get sufficient numerical records, he trained two native Africans, who sat outside the nests with stop-watch, notebook and pencil. They had previously watched at the nests of various song-birds, but found swift-watching far more of a strain, since nothing would happen for several hours and then it took place at about 70 feet per second. These African swifts, like our own, feed their young with large meals at infrequent intervals. Their other behaviour is also similar. So is that of the African palm swift (*Cypsiurus*), except that the young are fed rather more often. One of the observers had the luck to see the hatching of a young palm swift. The nest, it will be remembered, is a small pad with a narrow shelf to which the eggs are glued, and there is a sheer drop below, so that hatching is hazardous. As the nestling comes from the shell, the parent holds it in by pressing with its body on the palm leaf. Afterwards, the nestling grips the rim of the nest with its strong claws, on which its life depends, since if it let go, it would fall to the ground.

Few observations have been made on American swifts, and these mainly bear out what is known of other species. But there is a curious complication in the chimney swift, in that sometimes a third adult, and occasionally a fourth, joins with the breeding pair to help in brooding the eggs and feeding the young. Such extra helpers have usually been males but sometimes females. In three instances where there were two extra helpers and the sex of both was proven, both were males. The reason for the habit is not known. Sometimes the helpers have been yearlings, which doubtless do not breed, but at other times they have been older birds and there is no reason to think that they would be incapable of breeding (provided they could find a nesting site). Among British birds, an extra helper has sometimes been seen at nests of the long-tailed tit, and the same has been recorded of several American species, but the habit is very unusual.

Chapter 8

THE FEATHERED NESTLING

The feathers of the young swift grow through its skin like black prickles, and for a time it looks more hideous than ever (Plates 11, 12). When fully feathered, it is mainly black except for the pale throat, but it can be told from an adult because many of the feathers, including those of the wing, have white edges, an area above the bill is also white, and the throat is whiter than in the adult. These points can be seen in Plates 13 and 14.

Nestling robins, however much or little food each receives, usually grow their feathers at a nearly constant rate and leave the nest at a fixed age. As a result, the members of one brood leave more or less together, which helps the parents to keep charge of them. A disadvantage is that, in order to maintain their growth, any young which have received less food than the rest may be badly below weight and may be too weak to survive long. In the swift, matters are ordered differently. When the young fly from the nest they are independent, so there is no need for the parents to keep them together, and in fact members of one brood often leave on different days. Further, the food supply for the young is so uncertain that, if they had to develop at as steady a rate as young robins, either they would be unable to keep up in bad weather and would die, or too much food would be available in fine weather, which would be a waste. Instead, the rate of growth varies with the amount of food received. This is another valuable adaptation of the nestling swift, since in bad weather when food is scarce the bird can slow down its growth, thus saving energy for the vital functions.

The growth of a young swift is most easily measured by the length of the wing from its tip to the main angle at the carpus (or wrist).

The growth curves for three extreme individuals are shown in the bottom right-hand corner of Fig. 12. Bird A grew up during good weather and its wing reached a length of 100 mm (4 inches) when it was just over three weeks old. Bird C, on the other hand, had bad weather for most of its early life, and did not attain the same length of wing until it was nearly six weeks old. After this, the weather turned fine, and from then on it grew as rapidly as bird A. In contrast again, bird B was hatched in good weather and for the first 3½ weeks grew as fast as bird A, but bad weather then set in, and thereafter it grew much more slowly.

As a result of such differences, the nestling period of the swift varies by up to three weeks, whereas in song-birds of similar size it varies by a very few days. In good or moderate weather at Oxford, young swifts have left the nest at an average age of six weeks, the shortest period being 37 days, while a period of 35 days has been recorded in Germany. In bad weather the nestling period is prolonged by a week or more, and in each of the bad summers of 1948 and 1954, one of the nestlings stayed for as long as 56 days. The limits of the nestling period are therefore from exactly five to exactly eight weeks, a wide spread.

For its size a swift spends an unusually long time in the nest. A blackbird, for instance, remains only a fortnight. But the comparison is misleading, since after it leaves, a young blackbird is fed and cared for by its parents for another three weeks, whereas a young swift is independent and has to fend for itself.

Swifts have sometimes been recorded leaving the nest before they are five weeks old, but we think that in all such cases the bird was not yet fully grown. Occasionally, a nestling falls out of the nest when frightened, while if it is starved near the end of its time it often leaves prematurely. Such premature fledglings, which are specially frequent in bad summers, can be recognised by their abnormally short wings. Some of the young which leave in this way can fly at least a short way, and premature departure probably gives them a better chance of survival than if they wait in the nest for a meal that never comes. Incidentally, it may be such birds that have given rise to the notion that a swift cannot take wing from flat ground. An adult swift can do so, as we and others have tested, but starving young cannot, and they are the only swifts that one ordinarily gets the chance to test.

Before the feathers appear, the body of a young swift keeps at about the same temperature as its surroundings. Usually, therefore, it is as warm as the brooding parent, and when it is left unbrooded it cools down, as mentioned in the last chapter. When it has grown its feathers, it keeps a nearly constant warm temperature, like other feathered birds. In 1947, we took the temperatures of many feathered nestlings and of several adult swifts, by inserting a clinical thermometer in the cloaca. The adults registered 105° to 106° Fahrenheit, six or seven degrees higher than a man, as is normal in wild birds. The feathered young were cooler, 101°F to 102°F, occasionally only 99°F, but rising to 105°F shortly before they left the nest. The Finnish zoologist J. Koskimies took many accurate readings with a thermo-couple, with a similar answer.

Koskimies also took both young and adult swifts from their nests and recorded their temperatures (and other activities) as they starved to death. He found that at night, after several days of starvation, the young lost control of their body temperature, which fell nearly to that of their surroundings, sometimes dropping as low as 70°F. Nevertheless, they resumed temperature control on the following day, lost it again the next night, and so on. This must be reckoned another, and perhaps the most remarkable, of the adaptations of the young swift which helps it to tide over periods of food shortage. To maintain a high body temperature requires energy, and if the nestling can dispense with it, this energy is saved for more vital functions and hence life may be prolonged – a fascinating fact, though brutally discovered. Through losing their temperature control the young become torpid, which does not matter at night, but they need to resume their temperature control by day so as to be active enough to take food from the parents. In nature, of course, the result would have been less drastic than in Koskimies' experiments, since at night the parents would have been covering their young, which would thus have remained much warmer. It should perhaps be added that although these experiments seem cruel, they were not so bad as some of those that are performed on mammals in the laboratory, experiments that have led to great advances in theoretical biology and in the practice of healing. When we recover from illness, or are prevented from illness by inoculation, we should recall the price that has been paid.

Only one other kind of bird is known to lose its temperature control and to become torpid each night. This is the hummingbird. The smaller the animal, the greater in proportion is the surface over which heat can be lost, and hummingbirds are so small that they might lose too much energy at night if they maintained their high temperature. Hence they cool down at night, becoming torpid, and resume temperature control again in the morning. This happens regularly when hummingbirds are in full health and is not, as in nestling swifts, an adaptation to periods of bad weather.

Swifts are rather inefficient in keeping their nesting holes clean. When newly hatched, the young defaecate over the rim of the nest into the box. Later, after they are three to four weeks old, they often go to the entrance, turn round with the cloaca over the hole and defaecate outside, and occupied swifts' nests can be discovered late in the summer by the white splashes on the walls beneath them. But older young by no means always go to the entrance and often defaecate on the floor of the box. The parents tend to take away moist faeces, probably swallowing them when small and carrying them in the throat when larger. But many faeces get left behind, and when they have dried they are no longer removed, so that near the end of the season the nests get extremely dirty. Song-birds are much more efficient at nest-sanitation, helped by a valuable adaptation whereby the faeces of the nestling are produced in a gelatinous sac. This enables the parents to carry away the faeces unbroken. One wonders whether song-birds have evolved greater efficiency, partly at least, not from the desirability of cleanliness but because white splashes beside the nest would reveal it to enemies, a danger which does not apply to swifts in the safety of their holes. African palm swifts, unlike most other species, have open nests, but keep them clean because the young cling vertically to the rim of the nest, with the cloaca projecting over its edge, so that the faeces fall to the ground.

Newly hatched swifts stay in the nest, and if by chance they fall out, they cannot climb back and are ignored by their parents. But after they are about ten days old the young can climb back, and from the age of two or three weeks they shuffle about the box. At times they take definite exercises, flapping their wings violently and jumping up and down for several seconds at a time. Sometimes the body is tipped forward, the wings and tail are spread and the wings are vibrated. Starting when they are about a month old, the young also do 'press-ups', the wings being partly extended and pressed down on the floor, while the body is raised until both it and the feet leave the ground, so that the whole weight is taken by the wings. At first the bird cannot sustain itself in this curious position and merely hops up and down, but after a few days it can hold its body clear of the ground for a second or two, with the wings curving down on each side. Its strength gradually increases until, a few days before it finally departs, it can hold the attitude for as long as ten seconds. Such exercises, as would be expected, are seen chiefly in well-nourished young and rarely take place when they are short of food, as in the wet summer of 1954.

The exercises presumably help to strengthen the wing muscles. Most other nestling birds, and notably young eagles and falcons, flap their wings violently for some days or even weeks before they leave, standing on the rim of the nest and, in the later stages, rising for a short way from it and returning. It has recently been found that young nightjars do the same, not during the day, when they might be conspicuous to enemies, but only in the twilight, when they are comparatively safe. Young swifts, concealed in their holes, can exercise their wings without exposing themselves to attack, but the nest-site is so small that the exercises must take a peculiar form. In like manner the human 'press-up' provides for the flat-dweller the muscular exercise that his ancestors took in the forest. There is another human parallel, since when it is a few months old, but before it crawls, a baby spends much time lying on its back and kicking hard with its legs. There is a seeming compulsion about these movements, just as there is in the wing movements of a young swift, and in the baby also, they presumably help to strengthen the muscles before they are needed for progression.

A German zoologist confined some nestling pigeons in tubes so they could not flap their wings, though they could still be fed. He did this in order to test whether flight is a purely instinctive action, or whether it needs practise before it can be performed efficiently. When the birds were released, their flight movements and muscular co-ordination were as good as those of pigeons raised in the normal way, showing that flight is instinctive. But his findings have sometimes been taken to mean that the wing exercises of young birds are valueless and due merely to the premature development of an instinctive urge before the time that it can be fulfilled. Against this latter view, the wing exercises of young swifts are not just incipient flying movements but are specialised, which suggests that they have been specially evolved. Those of other young birds also differ from true flight, though not to the same extent as in swifts. Further, young swifts curtail their exercises when food is short, presumably in order to save energy. But even at ordinary times the young cannot afford to use energy unnecessarily, which suggests that the movements are of value. Finally, although young pigeons prevented from flapping their wings could later fly for a short way, their wing muscles had in some cases deteriorated. Hence, while the wing exercises of nestling birds perhaps originated through premature development of the flight instinct, their persistence and specialisation is probably due to the advantage of strengthening the muscles before the young leave the nest.

Eventually the time comes for the nestling swift to depart. Until now, its only direct knowledge of the outside world has been what it can see by looking out of the entrance hole. Afterwards it will not return to its nest, it will have nothing more to do with its parents, and it will fend entirely for itself. For the few days before it leaves, the young bird spends most of the day at the entrance hole. We have several times seen the actual departure. On one occasion, the feeding parent left the box at 8.25 a.m. The nestling sat looking out of the entrance, then partly spread its wings and tail and tipped forward with its head out of the hole, but turned back into the box. This performance it repeated. Finally, at 8.35 it put out its right wing from the hole and gently slipped out. In another box, a nestling was found at 8.10 a.m. sitting with its head out of the hole and its wings and tail partly spread. It then drew back, stretched its wings and preened them, moved back from the hole, then up to it again. The body feathers were alternately fluffed out and flattened, as if the bird were breathing deeply, and the tail was moved up and down. It then stretched its wings above its back and tipped forward half out of the hole, but 'feared to launch away' and scrambled awkwardly back into the box, supporting itself on its wings. After sitting at the entrance for another minute, it went quietly out. The preliminary hesitations are characteristic and have in some instances lasted for several hours or even for more than a day.

Nearly all of the young observed by us have left in the morning, often before eight o'clock, and this was also the experience of R. E. Moreau in two related African species. A few of our birds went in the afternoon or evening, but some of these may have been short of food, or disturbed. Other nestlings in the brood appear to take no notice of the departing bird and they usually leave on a later day. Also, all the young studied by us have left when their parents were out hunting.

Hence when the last of its young leaves, the parent bird is unaware that its nest will be empty until it returns on its next visit with food. It then acts in a disturbed way, moving on and off the nest several times, poking it with its bill, even displaying at it with one wing raised, also walking high on its feet as in threat display, or stretching its wings and gaping. These are presumably displacement reactions, as the bird is thwarted in its urge to feed the young. After this, it usually sits about half-way up the box facing the entrance and then either makes violent swallowing movements and leaves soon afterwards, or leaves with the food-ball still in position. The parents usually return to roost in the box for several further days, but their fledglings are not seen again. Presumably, they set off at once, or almost at once, on their journey to Africa. There is perhaps no more striking instance in birds of the efficiency of inborn behaviour than this of the young swift, which after its cramped life in the nest enters on its aerial life in a very wide world, feeds itself, finds some way of spending the night, and migrates, all without help or guidance from its parents.

Chapter 9

FEEDING HABITS

The remarkable adaptations of the nestling swift are necessitated by the feeding habits of the adults. With their long thin wings, huge gape (see Plate 17) and short bill, swifts are highly specialised for catching insects in flight in the open. They take them in the mouth but do not, of course, like aerial leviathans, just fly with their mouths open, filtering off those insects which happen to enter. Instead, they can diverge somewhat from a straight course to take insects seen to one side or the other, and do so very rapidly when need arises. They also select their prey to some extent from among the insects that they meet, and like other birds they seek places where their food is abundant and avoid places where it is scarce.

The places where air-borne insects are abundant, and where swifts hunt for them, depend to some extent on the weather. On fine summer days swifts often feed from twenty to a hundred feet above the ground, sometimes swooping lower in the open where there are no obstacles to endanger them. If there is a fresh breeze, they often hunt up and down on the windward side near the top of a belt of trees at right angles to the wind. Here there is an upcurrent, which may help them to maintain height, while the wind probably concentrates the insects from off adjoining fields. On fine and still days swifts often circle up to several hundred and occasionally to several thousand feet, but since in fine weather, as in bad, insects are most numerous close to the ground, it is uncertain whether they are seriously feeding at such times.

If there is a lake near the nesting colony, as at Woodstock near Oxford, the swifts regularly hunt over the water, to which they can safely approach within a few feet without danger. When a wind is blowing across the lake, they often patrol along a belt of trees on the windward side, in the same way that they patrol a belt of trees beside a field, and swallows and martins often join them there. On large meres swifts appear particularly in windy and wet weather, when land living insects tend to stay on the ground, but some of those with aquatic larvae fly low over the water.

Swifts are likely to appear anywhere where insects are numerous. Thus thousands may gather when midges are rising in huge numbers off a mere or sewage farm. More unexpectedly, during a plague of the pine beauty moth in one of the new state forests of the Midlands, swifts flew in hundreds up and down the rides, though normally there would be none in these dark and regimented pine woods. Parent swifts tend to feed near their nests. Indeed, if they did not do so, there would seem little point in their breeding in small dispersed colonies. Some swifts can usually be seen feeding close to the tower, and if there is sudden rain many enter their holes at once, suggesting that they had not far to come for shelter. Hence the large gatherings of swifts over water may consist chiefly of non-breeding birds, especially yearlings, though this is not proven. That breeding swifts cannot afford to travel too far when feeding their young is suggested by their disappearance as a breeding species from central London. With the increase in houses and smoke, insects have become much scarcer than formerly in the centre of the city, and feeding journeys to the suburbs would presumably take the swifts too long.

The food of most birds can be studied only by killing them and analysing what is found in their gizzards. This is unsatisfactory because, apart from the destruction of the birds, insect fragments are often hard to identify, while the proportions present in the gizzard may not give a true picture of the diet, since some foods are digested much more quickly than others. Thus a recent study of the food of the rook by gizzard analysis suggested that earthworms were unimportant; but they are digested rapidly and digestion may continue in a shot rook after death. In fact, observations on the foods brought to young rooks showed that earthworms were the main item.

Fortunately, members of the Edward Grey Institute at Oxford have found a way of studying the food of swifts that is easier, more reliable and more humane than killing and dissecting the birds. If a young swift is taken from the nest just after it has been fed by its parents, which is easy in the tower, the food can readily be manipulated from the throat before it has been swallowed. This does not hurt the nestling, and all that it loses is a single meal at extremely infrequent intervals. If the returning parent swift is caught before it has fed the chick, it likewise may disgorge the meal, but this is not to be recommended, as parents sometimes desert their young if caught in this way.

The swift usually brings just over one gram of insects to its young in each meal. The meals are of the same average weight in fine weather, when insects are plentiful, as in wet weather when insects are scarce, the difference being that the parents take much longer to collect a meal in wet weather. We sometimes measured the speed of collection precisely, one parent (on a fine day) collecting 1.7 grams of insects in 64 minutes and another 1-2 grams of insects in 47 minutes. The largest recorded number of meals brought to one brood in a day was 42, so that the parents probably took about 50 grams (1¾ ounces) of insects for their young, in addition to whatever they ate themselves. Each meal usually contains between 300 and 1,000 small insects and spiders, but a few meals with larger insects have contained only 100, while others with very small insects have had up to 1,500 individuals. The swifts which collected 42 meals in a day probably removed about 20,000 insects from the air in so doing.

Swifts catch a huge variety of insects, so that precise identification of even a few meals may take the entomologists several years. From only twelve meals collected at the village nests near Oxford, 312 species of insects and spiders were certainly identified, but various difficult specimens could not be identified, and no one could be found willing to identify the numerous Hymenoptera and the few small moths. Over 400 species may well have been represented. Moreover, this sample included none taken over water. The swift probably eats more species of animals than any other British bird. This is because most birds seek their prey in a restricted type of country, but insects from almost every habitat find their way into the air and so into the mouths of swifts.

The commonest insects in the meals have been dipterous flies and plant-bugs, especially aphids and their allies. Hymenoptera, beetles and spiders were also numerous, and there were a few specimens of most other insect orders, including mayflies (especially in one meal on a wet and windy day), small dragonflies and small moths. There were great variations from day to day in the species caught, depending on the weather, on where the birds were feeding and on the seasonal cycle in insect life. All the insects could have been taken on the wing, and though there are two published records of swifts alighting under the eaves of houses and pecking insects off the walls, this must be quite exceptional.

There was a pleasing incident in connection with the sample meals on one July morning of continuous rain. We visited the tower about noon as we wanted further samples to see what the swifts brought for their young in wet weather. About half an hour after our arrival, a swift entered to feed its chicks and two others did so soon afterwards. We obtained the meals in the usual way. Then several other swifts entered and fed their young, so many visits in a short time being most unusual. To our further surprise, a quick look showed that two of the three meals contained many hoverflies, insects that are typically on the wing only in warm and sunny weather. When we left the tower we found the reason. The rain had cleared away and it was warm and sunny. We presume that, on this day, many of the parent swifts had flown off from the tower westwards through the rain-belt, meeting the fine weather (of which we ourselves had no knowledge) and then travelling back with it to Oxford, feeding as they came. This explains why so many arrived back at about the same time and how they came to take the hoverflies.

In recent years nets have been suspended from towers, kites and barrage balloons in order to study the insects drifting in the air. The work was started by Professor A. C. Hardy (before he came to Oxford to be head of the department which includes the Edward Grey Institute), and as an unexpected result of it he saved the spindle tree for Britain. The harmful bean aphids feeds for part of its life-cycle on spindle, and at the start of the war an order was nearly issued for the destruction of all spindle trees, in the belief that this would rid us of the pest. But aerial tow-nets showed that each year large numbers of bean aphids drift into England on the wind from the Continent, so that the measure would have been useless. Pure research undertaken for its own sake may have unlooked-for practical value.

The work with aerial tow-nets has shown that insects are most numerous in the air on fine, warm and still days and much scarcer on wet, cold or windy days. It is presumably for this reason that, as shown in Fig. 13, swifts rarely breed on the western coasts of Britain, for here the weather is usually wet or windy; also the prevailing westerly winds, coming off the sea, carry no insects with them.

Observations mentioned later show that nestling swifts are fed at much more frequent intervals in fine than in bad weather. A precise analysis of the influence of the weather on the growth of the young was made in two years, in which every young swift in the tower was weighed once a day. From these figures we first made a curve for the average increase in weight to be expected at each age. The actual change in weight of each nestling on each day was then compared with the expected change for a nestling of the age concerned, and

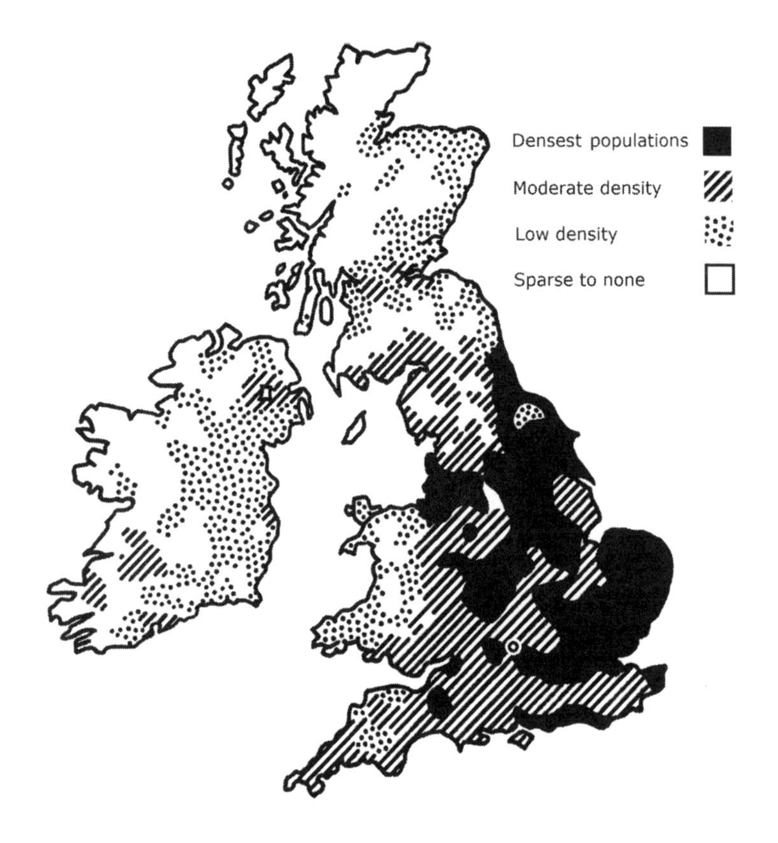

Fig 13: Breeding range of the swift in the British Isles. O=Oxford

the difference between the observed and expected figures was summed for all the young on each day. The total so obtained was then compared with the sunshine, temperature, rainfall and wind on the day in question. The analysis was laborious, and it was complicated because sunny days tend also to be warm and calm, and wet days to be cold and windy, but the influence of each element in the weather could be assessed separately with the help of a statistician and a calculating machine. The results showed clearly that nestling swifts grow better on sunny than dull days, on warm than cold days, on fine than wet days and on calm than windy days. Needless to say, the analysis took far longer than the time needed to write this short paragraph about it.

It is interesting to compare the insects taken in aerial tow-nets with those caught by the swift. The figures show that a swift catches prey much more quickly than would a moving net of the same size as its gape, as was to be expected since a swift seeks places where prey is abundant and can turn somewhat off its line of flight to catch an insect. Further, the swift selects its prey in respect to size, rarely taking insects shorter than 2 mm or longer than 10 mm (about ½ inch). An instance of selection at the upper limit of size was once seen near the tower in late June. Many dor-beetles of the species *Amphimallus solstitialis*, which are larger than the insects usually taken by swifts, were flying up from the grass and collecting round trees. Swifts often swooped at them and sometimes caught them in the air, but then dropped them again.

It was to be expected that swifts would avoid insects that are too large for them to manage and that extremely small ones would not be taken, but a more subtle selection for size was revealed by the analysis of the meals. In fine weather the meals consisted chiefly of insects 5 to 8 mm long, but in poor weather 2 to 5 mm long. Now insects 2 to 5 mm long are more common in fine than in poor weather and are more numerous than larger insects in all types of weather. Presumably, in fine weather when larger insects are plentiful, swifts can collect a meal more quickly if they do not go out of their way to catch the smaller kinds. In bad weather, on the other hand, larger insects are so scarce that the swifts cannot afford to be so selective. The only small insects sometimes taken in large numbers in fine weather were aphids, but these are at times so abundant that it would be rewarding to hunt for them.

A parallel example of selection of prey by size has been found in the blue tit near Oxford. In June, blue tits feed chiefly on leaf-eating caterpillars but in July on aphids. Now aphids are already common on the leaves in June, but blue tits do not spend time hunting for them then, but only later, after large caterpillars have become scarce.

A British observer noticed for several years that swifts regularly hunted round his bee-hives. Thinking that they must catch a great many bees, he at intervals shot a swift, eventually collecting eight. When he dissected them, he found that they had eaten only the drones, which are, of course, both harmless to the swift as they are stingless and valueless to the bee-keeper as they do not make honey. If, as seems likely, the swift hunts at a speed of about 25 miles per hour, it can see its prey only a short time before catching it, and it is remarkable under these circumstances that it can distinguish drones from worker bees. The incident also shows how careful one must be in deciding whether a bird is harmful.

The same ability to recognise drones has been found in other birds which sometimes take bees, including the alpine swift, European swallow, American cliff swallow and European bee-eater. The bee-eater has often been shot as a pest in countries where it is common; but how many have been examined to see what they were really eating? A recent study in Czechoslovakia has shown that, despite its name, the bee-eater feeds chiefly on dragonflies and butterflies, while hive-bees formed only two per cent of its diet and all of them were drones.

Although only drone bees were found in the gizzard of an alpine swift, another had eleven bee stings in its throat. Perhaps this individual had not yet learnt to distinguish drones, but there is another possibility, that only drones not worker bees are found in birds' gizzards because the workers sting the birds and so are dropped not swallowed. Against this idea, the common swifts mentioned earlier were regularly seen hunting round bee-hives and would scarcely have done so if often stung. Also the observer who saw American cliff swallows pursuing bees noted that they took a long time over each capture, as if making sure that it was a drone. It is, of course, possible to make the distinction. Thus I have seen Oxford's professor of entomology, in his earlier days, collecting hornets in flight with his bare hands. He, of course, could distinguish the stingless males, but it was impressive to see him release six hornets from his cupped hands in the office of the war-time organisation for which we at that time worked.

The swift is one of the latest of small diurnal birds to leave in the morning. This is probably because its food is scarce during the first hours of the day. Aerial tow-nets have shown that air-borne insects are most numerous during the afternoon. Correspondingly, our counts at the tower have shown that in poor weather the parents feed their young most frequently between 4 and 5 p.m. GMT. In good weather, on the other hand, feeding is most frequent in the hour before midday. The difference in fine weather, it may be suggested, is that the young have already received much food before midday, so they beg less vigorously from their parents, which then come less often, presumably taking time off to feed themselves. In poor weather, on the other hand, the young continue begging vigorously, so that the parents hunt actively throughout the day. But even in bad weather there is a slight reduction in their effort around noon, which perhaps means that, whatever the weather, the parents seek some food for themselves at this time of day.

Two Continental workers have made regular observations on the time of day at which swifts leave their nesting holes in the morning and return to them at dusk. In Germany at latitude 50° N, they depart on the average fifteen minutes before sunrise. One to two degrees further south they leave five minutes later in relation to sunrise, and one to two degrees further north five minutes earlier. In latitude 53–54° N, they leave half an hour before sunrise and in

Finland at latitude 60–63° N a whole hour before sunrise. There are corresponding variations in the time of retiring to roost in relation to sunset. These differences can readily be explained by supposing that swifts emerge in the morning and retire at dusk at a particular light intensity, since in summer the twilight lasts longer the higher the latitude. Moreover swifts leave earlier on fine than dull days.

Because of these variations, swifts have a much longer working day at higher latitudes. Thus around Helsinki at midsummer they stay in their holes for only about four hours each night, and in Swedish Lapland north of the Arctic circle, where the daylight is continuous at midsummer, we saw them return as late as 10.20 p.m. on a misty evening. Swifts have been seen flying at midnight in the Arctic, but as they may stay out all night in the dark at lower latitudes, this has no special significance. The general view is that, even where the daylight is continuous, they normally come in for about three hours each night.

At the tower we found that some individuals came in to roost consistently earlier than others. The earliest to come in were among the earliest each evening, and the latest were nearly always among the last. Apart from these individual differences, in the autumn parent birds still with young almost always came in to roost much later, sometimes twenty minutes later, than those which had not bred or whose young had already gone. Those with young to feed evidently continue hunting until the last possible moment.

The food of other species of swifts has been very little studied. All feed in the same manner as our bird on insects in the air, and it is reasonable to suppose that the larger species of swifts tend to take larger insects and the smaller species of swifts smaller insects. Actually, the various swifts do not differ greatly in size, the largest species having a wing-span of between two and three times that of the smallest. A swift's wing is adapted for high speed but not for rapid and precise manoeuvre, and it seems possible that, below the lower limit of size, this design is not that best fitted for catching tiny insects. Anyway, at about this limit of size the swallow tribe takes over. Swallows have a shorter and less specialised wing and travel less rapidly than swifts, but they have a greater ability to check and turn in flight, so that they can hunt freely for very small insects close to trees and other obstacles where swifts cannot come. At the other end of the scale, it seems likely that if swifts were any larger they would be unable to find enough large day-flying insects to support themselves throughout the year, since most air-borne insects are small. At night the situation is different, as many moths and other large insects then take wing. But swifts, like most other birds, cannot see well enough to catch insects in a dim light, which indeed may be why so many large insects come out at dusk. By doing so they escape many enemies, but not all, since here the nightjar tribe takes over, many of which are larger than the largest swifts.

There is thus an economy in nature, each source of food being exploited by a particular type of bird which is appropriately adapted. This idea found the somewhat absurd expression in an anonymous Victorian book for children. 'Why,' asks the child, 'are swifts and swallows appointed to their different ranges of elevation when on the wing?' and the answer comes, 'Very high in the atmosphere the fewest flies are to be found: there the rapid swift is stationed. The swallows take a lower region and fly more slowly, for in their appointed track there are more insects to be met with. Thus their united efforts clear the air of insects which would otherwise be injurious to vegetation and annoying to animals.' The basic observation is mainly correct, but the inferences drawn from it show how far our great-grandparents had to travel before they could accept the idea of animal evolution.

Chapter 10

FLIGHT

It is time to consider the flight of the swift, for on that so much else in its life depends. The bird-books written earlier than a century ago started with the eagle, for he was the King of Birds. But about the time when monarchs were tumbled from their thrones by black-coated republican intellectuals, the bird-books followed suit, banishing the eagles to a back page as primitive and giving leadership to the crows. In the era of the common man the order has been changed again and now reaches its climax with the sparrows and finches. But perhaps with an age to be dominated by air power it would be fitting to change once more, and to give pride of place to the bird which excels where birds as a group excel, in flight. For this honour the swift is the strongest candidate, for it is perhaps the fastest of all birds in straight flight and certainly the most dependent of all on the air.

> Like a rushing comet sable
> Swings the wide-winged screaming swift.

So wrote Lord de Tabley, while among modern poets Walter de la Mare has spoken of their 'wild-winged archery' and Laurence Whistler of 'the power-dive on to the thatch', a phrase which recalls our own early days with swifts in the village near Oxford. But except for these brief mentions, the mastery shown by a swift in the air has not received the admiration that it deserves. The analysis of its flight is a technical subject which I have not studied at first hand and what follows is a summary of the work of others in so far as I understand it.

A flying bird has two problems, to keep up and to move forward, the latter normally needing the greater effort. When air passes over an object cambered in cross-section like a wing, a reaction is set up giving a force mainly upward (lift) and slightly backward (drag). Further force is needed if the winged object is both to keep up and to move forward, and this is supplied in aircraft by propellers and in birds by flapping the wings. Men in gliders and slow-soaring birds like vultures can, however, get the needed force from upcurrents, especially those caused by wind against steep slopes or by hot air rising from rocks or sand heated by the sun. In hot countries glider pilots seeking upcurrents often find themselves in company with eagles and vultures doing the same. Such soarers have wings that are extremely large in proportion to the body and of such a shape that they fall through the air as slowly as possible. They move forward by losing height relative to the air, falling gently forward as it were, which they can afford to do when the air is rising at least as fast as they are sinking through it. To give a big lift a broad wing is needed, which means that the drag is also great, and the bird moves forward slowly, but that is how it seeks its prey, gliding at a height and scanning the ground below. Vultures, it may be added, are more efficient than human gliders because, in an emergency when air currents are failing, they can flap their wings to help them out.

A quite different type of gliding or sailing is found in albatrosses and to some extent in gulls, which have long narrow wings. In a narrow wing the drag is reduced, hence the bird can fly faster. The lift is also reduced, but since more lift is given when air flows faster past the wing, the bird can stay up so long as it is travelling fast. At sea in a strong wind, the air is moving more slowly close to the sea than higher up. An albatross makes use of this by flying down-wind in a long glide, losing height as slowly as possible; then, when it is near the surface of the sea, it turns and rises steeply into the wind, which brings it into an increasingly faster stream of air, hence its airspeed remains high though its speed relative to the sea is low. Provided that the wind is strong enough to give a big reduction in the speed of the air close to the water, an albatross can maintain itself in this way without flapping. That is why albatrosses are found chiefly in regions of strong winds, notably the Roaring Forties, and the only one found in the Doldrums, the Galapagos albatross, often flaps its wings, which look broader than those of the other species.

Both the slow-soaring vultures and the fast-sailing albatrosses fly for long periods without beating their wings. At the other extreme, various birds beat their wings almost continuously when flying and rarely, if ever, glide. The swift is intermediate, as it usually beats the wings a few times in rapid succession, then glides for a few seconds, and then flaps again. Its narrow wings enable it to glide fast, but it cannot go far without losing height, so that each glide lasts only a short time. Many kinds of birds glide with the wings held rather above the horizontal, but in swifts, as in certain petrels, the wings may be held a little below the horizontal, a position which gives greater speed and manoeuvrability but less stability. But stability is far less important in birds than aircraft, since when the need arises a bird can react much more quickly than a pilot.

Swifts, like most other birds, use upcurrents when they are available. On a windy day in a town they often utilise the rising gusts caused by the wind beating against walls, and they gently lift a wing to clear a roof, like a shearwater clearing a wave at sea. Swifts also hunt for food in the upcurrent on the windward side of a belt of trees, as already mentioned, and at times they rise in the updraught of an approaching thunderstorm. With a following wind on migration they may travel by drifting in circles, which again indicates the use of upcurrents. Under all these conditions, however, they alternate short glides with rapid wing-beats. They are not sustained gliders.

It is unfortunate that the same word 'wing' is applied to both bird and aircraft, since a bird's wing gives propulsion as well as lift, and so is equivalent in function to both the wing and the propeller of a flying machine. In flapping, the chief part is played by the outermost feathers, the primaries, as these are furthest from the body and so exert the greatest force as the wing moves downward. They are attached to the bones of the wrist and hand, which for greater strength are fused together, not separate as in man. Lift is provided chiefly by the inner feathers, the secondaries, which are closer to the body and are attached to the forearm. A song-bird or pigeon has what might be termed a general-purpose flapping wing, the upper arm (humerus), forearm (ulna and radius) and hand (carpus) are of similar length, while the secondaries (giving lift) and the primaries (giving forward propulsion) occupy areas of equal size on the outstretched wing. In birds of this type propulsion comes from the downstroke, while in level flight the upstroke is effectively wasted, being merely a way of getting the primaries back into position for the next downstroke. On the upstroke, the wing is partly folded at the wrist (carpal angle), and is turned edgeways on with the primaries partly open, little resistance is offered to the air and the wing moves up more or less of itself. Consequently the muscles used on the downstroke, the pectoralis major, are much larger than those used to raise the wing, the pectoralis minor, being in the ratio of nine to one in various song-birds. These flight muscles are attached to the keel of the breast-bone.

The details of the upstroke vary in different species, especially in the extent to which the wing is folded at the wrist, which is great in song-birds but less in long-winged birds such as gulls.

Also, the wing is raised more strongly at the moment of taking off than in level flight, and this is the time when the pectoralis minor muscles are chiefly used. A rising pigeon gives a rapid backward and upward flick of the wing-tip at the end of the upstroke, which produces an upward and forward force, but the bird can maintain this effort for only a few seconds at a time, and in level flight the upstroke is not powered.

A powered upstroke also occurs in hummingbirds when they are hovering. The body is nearly vertical, so that the bird makes two sidestrokes rather than an upstroke and a downstroke. Both strokes are powered, the action being like that of a paddle, and the wings are winnowed at up to 75 times a second, a frequency within the range of our hearing, hence the hum. On the upstroke the wing remains fully extended with the primary feathers locked together, so that the action is quite different from that in most other birds, which bend the wing at the carpal angle and open the primaries. The hummingbird's wing consists almost entirely of the primary feathers, which are supported on the relatively large carpal bones, while the middle arm is short and the upper arm bone is extremely short and squat, which helps in the rapid rotation of the wing in winnowing. The wing has a shallow camber, which reduces the drag, and also helps when the wing is turned over on the upstroke. The pectoralis minor muscles used on the upstroke are nearly half the size of the pectoralis major, proportionately much larger than in any other type of bird, and the short narrow wing is effectively all propeller, giving amazing control. Thus a hummingbird can hover in front of a flower to suck nectar, keeping its position so accurately that it seems hung on a wire, and can then shoot off sideways too rapidly for the eye to follow it. Hummingbirds are also capable of extended flight, since one species regularly migrates across the Gulf of Mexico on a nonstop flight of at least 600 miles. The wing action in such continuous flight has not been studied.

The swift has a long, narrow, swept-back wing with a shallow camber, clearly designed for high speed (Plate 3). John Barlee informed me that its 'coefficient of dynamic similarity', a term used by aircraft designers which takes into account the wingspan, wing-area and body-weight, works out at 2.4, the same figure as for a manx shearwater or a Meteor jet aircraft (Mk 8). It is much the smallest 'racing model' among birds, and since air resistance becomes

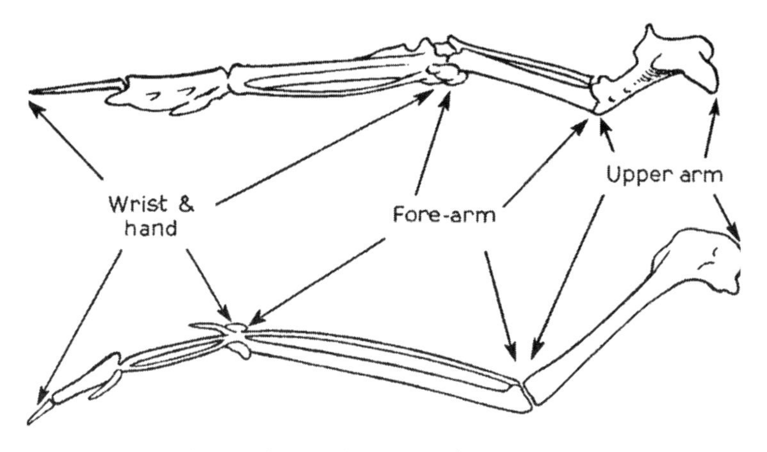

Fig 14: Wing bones: above, swift (just over life-size); below, chaffinch (1½ times life-size)

proportionately more important the smaller the bird, a swift is more finely designed for high speed than any other species. Its flickering wing-beats look peculiar and suggest that it may have an unusual wing-action. This is probably so, for though the long recurved wings are similar in shape to those of a falcon, in structure they closely resemble those of a hummingbird. Thus as in a hummingbird, the long primary feathers make up most of the wing, with a short area of secondaries close to the body, while the carpal bones are relatively large and the upper arm bone is short and squat. The bones are shown in Fig. 14, with those of a typical song-bird drawn to the same scale. As in a hummingbird, also, the primaries of the swift are stiff and grip tightly together, so that they cannot be separated by the bird, and when the wing is fully extended, the outermost primaries are bent slightly backward and the inner ones are bent slightly forward by the tension. The breast-bone has a remarkably deep keel for the attachment of the powerful flight muscles, which are dark red in colour indicating an efficient blood supply. In California, W. W.

Mayhew recently examined the bodies of a number of Vaux swifts that were drowned when roosting in a pipe, and he writes that the pectoralis major muscles are large for the size of the bird, while the pectoralis minor, instead of being narrow and strap-like, as in many birds, are relatively broad, about half the width of the sternal keel, and running along the keel for its full length. They are about one-fifth of the size of the pectoralis major, relatively much smaller than in a hummingbird, though larger than in most other birds.

Fig 15: Downstroke (left) and upstroke (right) of flying swift, with suggested resultant forces
(adapted from D. B. O. Savile)

Struck by the similarities between the wings of a swift and a hummingbird, an American worker has inferred that the upstroke of a swift, like that of a hummingbird, gives a forward force, as suggested in Fig. 15 based on his diagram. It is clear from its structure that, on the upstroke, a swift's wing is not appreciably bent and the primaries are not separated, and these points were confirmed by a short slow-motion film of flying swifts kindly lent by C. Eric Palmar, which also showed that the wingtips are remarkably far back along the body at the end of the downstroke. But extremely high-speed photography will be needed to establish whether the upstroke is truly powered. It might seem surprising that, if a swift has a powered upstroke, the pectoralis minor muscles should be only one-fifth of the size of the pectoralis major, since this ratio is much smaller than in a hummingbird and somewhat smaller than in a pheasant or pigeon, though larger than in most other birds. But a hummingbird hovers, while a pheasant and a pigeon use a powered upstroke to raise themselves rapidly from the ground. A swift, on the other hand, never hovers and in its normal life never has to rise rapidly from a stationary position. When this is borne in mind, its pectoralis minor muscles seem remarkably large, suggesting that they must have some other use, presumably in level flight, but further research is needed to settle the matter.

When flying into a strong wind, a swift usually flaps steadily and continuously, as otherwise it would not make sufficient forward progress, but more usually it alternates vigorous flaps with glides. The latter method takes it more slowly through the air, but is probably that which enables it to travel furthest for the least expenditure of energy. In this connection, J. Barlee informs me that, in a competition in America to see what type of car could travel furthest on a gallon of petrol, the winner was not, as might have been supposed, a low-powered machine which travelled steadily, but a high-powered one with the wheels modified for efficient coasting, which raced up to an extremely high speed, then coasted with the engine off until the speed had fallen very low, then accelerated again, and so on. This method is efficient only if the car can accelerate rapidly and coast smoothly. The swift, likewise, is a high-powered model, capable of rapid acceleration and fast gliding.

Swifts are generally claimed to be the fastest of all birds in level flight, but it is hard to be sure, as they rarely fly as rapidly as they can, and hardly ever give one the chance to make an accurate measurement over a straight course. A Dutch worker has estimated that, when feeding, swifts often fly at only 25 miles an hour, being passed by starling or pigeon, but that they sometimes move at 40 miles an hour and may increase up to 60 miles an hour if extended. A pilot in the First World War thought from his air-speed indicator that swifts in unhurried flight near to him were moving at 68 miles an hour. The swift's highest speed on a short burst has never been measured. One taken from a church tower at Tournai in Belgium and transported by aircraft returned from London airport in four hours, at an average speed of 37 miles an hour, a remarkable achievement even if it flew directly back, and also showing that the bird can orientate itself with accuracy.[1]

Fast flight depends on a smooth flow of air over the wings and can be greatly retarded by turbulence and eddies. Hence the surface of the wing in both aircraft and bird is extremely smooth, and to keep it smooth the swift spends much of its time in the nesting hole preening itself. Many birds also have devices to reduce the eddies set up at the edge of the wing. In eagles and vultures the wings are slotted, the inner side of each of the outermost primaries being narrower at the tip than the base so that, by spreading the wing, the bird produces a series of gaps near the wing-tips. The alula, the small group of feathers attached to the thumb-joint, may also act as a wing-slot. Such devices are needed particularly for a bird flying slowly, as when soaring, or when it is climbing steeply and the wings are held at a steep angle, as in a pheasant. They are not so necessary in, and would retard, a narrow-winged fast-flying bird like a swift, but in a

[1] A speed of 200 miles an hour has been claimed for the needle-tailed swift. In Assam, E. C. Stuart Baker timed the birds passing over his bungalow to a ridge two miles away, over which they seemed to dip. If, as seems possible, he lost them to sight against the background of the hill before they had covered the full distance, his records must be rejected. Incidentally, this large spine-tail has twice wandered to England, rashly since it thereby achieved posthumous fame as a stuffed exhibit.

swift the airflow is assisted by the taper of the wings and by their being somewhat recurved, with a 'fairing' of feathers to make the join between the trailing edge of the wing and the body less abrupt.

A bird has this advantage over an aircraft, that it can somewhat alter the shape of its surfaces in flight, modifying its aerodynamic properties. Thus when a swift checks in flight, it can turn the wing more broadside on and spread its tail, giving greater control, as shown in Plates 4 and 16. A swift can also bend the wing at the wrist, bringing the primaries closer to the body, as when gliding down towards its nesting hole.

The swift's long narrow wings prevent it from rising quickly and steeply from a flat surface. Some fast-flying narrow-winged birds, such as the golden plover or a duck, take off after a rapid run along the ground or the surface of the water. The swift, with its short legs, cannot do this, which is why it nests and roosts with a drop below. It then launches into the air by dropping, which quickly gives it a high air-speed, from which it gets the necessary lift to continue. Contrary to popular belief, it is not impossible for a swift to take off from a flat surface, but it does so much more clumsily than most birds and gains height slowly. If it had to take off from flat ground in nature, it would probably be exposed to attack from hawks.

A song-bird can rise from the ground much more quickly than a swift, but the extreme development of this type of flight is found in game-birds like the partridge. Partridges spend almost their whole life on the ground and take wing only to escape from enemies such as foxes. Members of the ICI Game Research Station at Fordingbridge in Hampshire have watched a flock of partridges throughout two entire days, and found that they took wing only when frightened, and then for less than two minutes all told. There could be no greater contrast with a swift, both in behaviour and in the type of wing. The wing of the partridge is specialised for a steep and rapid climb, so is very short and broad, highly cambered, with well-developed secondaries and broad well-slotted primaries, while the pectoralis minor muscles are large, about one-third of the size of the pectoralis major, as can readily be seen when the bird is served for the table. This type of wing is admirably fitted for a rapid climb and a sudden burst of speed, but not for sustained flight, and in fact partridges rarely fly for more than a few hundred yards before alighting again. This is sufficient to stop pursuit by a fox.

Probably because it would be so slow in taking off if danger threatened, a swift does not, like most other birds, alight to drink or bathe. It drinks by descending through the air in a shallow glide to an open patch of smooth water, sipping as its head touches the surface, then at once rising again, shivering as it rises to remove any drops of water on the plumage. A captive swift that learnt to take water from its owner's lips made similar shivering movements, although they were not in this case needed.

Swifts probably use falling rain for bathing, and when they are flying slowly through falling rain they not infrequently shake their feathers, as after drinking. The feathers of a captive swift were said to look very glossy after it had been given artificial rain. In Java, a group of cave swiftlets regularly flew out from their nesting place when heavy rain fell, presumably in order to bathe. In dry weather, swifts have another way of bathing, which we have not seen ourselves, probably because long dry spells are so rare in England. Swiss observers have seen common swifts descending in a fast glide to the river Aar and submerging themselves almost completely before rising again, with vigorous shakes of the body to remove the water. The same behaviour has been seen in the American chimney swift. There is also a record, from England, of 'smoke-bathing', a party of swifts repeatedly flying through the smoke from boilers, but dispersing when the smoke ceased to rise. House martins were doing the same, and various song-birds have been seen to station themselves on top of smoking chimneys. One wonders what the instinctive basis of such behaviour could be, and also what its value may be to the birds.

A swift travels so fast that, if it hits an obstacle, it may seriously hurt itself. There are several published records of swifts hitting walls and falling stunned or dead to the ground, while another hit a man's hat in Dublin and fell senseless, and two others collided with each other in the air, both being killed. Two other swifts have been seen to avoid a head-on collision by spreading their wings to the fullest, gliding up and meeting breast to breast.

It may seem surprising that a swift should be stunned by striking a hat, even though the incident occurred in the last century when hats were harder than now. This effect was perhaps due, not to actual contact with the hat, but to the sudden reduction in the bird's speed, causing a rush of blood to its head. The danger of a sudden change in speed is well known to fighter pilots. With a sudden increase in speed, blood may fail to flow along the main arteries to the brain and blacking-out results, which seems to be harmless except for the temporary loss of sight. But the opposite effect due to sudden slowing down can be much more serious, as it greatly raises the blood pressure in the head, resulting in haemorrhage, permanent damage and even death. It has recently been shown that fast flying birds, including the swift, various hawks and waders, differ from most other birds in having thin, transparent, single layered 'windows' in the otherwise opaque and double layered skull. These thin areas, it is suggested, are capable of movement or expansion, thus reducing the effects of high blood pressure when the bird has to slow up suddenly. Similar windows are also found in various diving birds, such as the kingfisher and terns, which have their speed suddenly checked as they strike the water. Swifts also have unusually large and flexible eyelids, which are perhaps adapted to prevent damage to the eye from small objects met in flight at high speed, and as a further protection the eye is set deep in the head.

Bird-watchers have often claimed that both the European swift and the chimney swift of North America can beat their wings alternately. Aerodynamically this seems impossible, and one would suppose that any bird which tried it must fall. In fact, a cine-film taken twenty-five years ago gave no evidence for alternate beating, and recent observations with the help of a stroboscope have been to the same effect. There are times when a swift appears to beat its wings alternately, but this is an optical illusion. As already mentioned, a swift beats its wings too quickly for the individual strokes to be seen by the human eye. When the bird turns, one wing is beaten more strongly than the other, and an American worker has suggested that the appearance of alternate beating is given when a swift banks and turns somewhat off its course to catch an insect, then banks and turns again, the eye seeing only some and not all of the wing-beats involved in this manoeuvre.

For efficient streamlining, a body should taper towards a point at the back. The swift has a forked tail, but in fast flight this is closed, so that the body ends in a point. The tail is opened to show the fork chiefly when the bird is flying slowly or turning, as in Plate 4. The main use of the tail is not as rudder, but to give greater control at a low speed through the air. The degree of control is greater the wider the tail can be spread, but a broad tail increases the drag, so is disadvantageous for a fast-flying bird. Most other birds with strongly forked tails, such as swallows, terns, frigate-birds, kites and pratincoles, also resemble swifts in having long and narrow wings. I would suggest that a forked tail is advantageous because it can be spread widely when needed, but when closed brings the back of the body to a point.

The extent to which the tail is forked varies in different kinds of swifts. The common swift, for instance, has a deeper fork than the pallid swift (*Apus pallidus*) or African black swift (*A. barbatus*), while the house swift (*A. affinis*) has scarcely any fork and the African white-rumped swift (*A. caffer*) and mouse-coloured swift

(*A. myoptilus*) have acutely forked tails with the outermost feather emarginated on its inner side. These variations seem connected with the shape of the wing-tip. In the species with strongly forked tails the second primary feather is the longest, whereas in those with less forked tails the first primary is about as long as the second, so that the wing ends less sharply. This correlation also appears to hold in other genera of swifts, in which the spine-tails (*Chaetura*), some American black swifts (*Cypseloides*) and some cave swiflets (*Collocalia*) have square-ended tails, while palm swifts (*Cypsiurus* and *Tachornis*) and scissor-tailed swifts (*Panyptila*) have long and delicately forked tails, and also more pointed wings. The reason for the correlation is presumably aerodynamic, but has not been studied. It may be added that even those swifts with acutely forked tails carry them folded to a point in normal flight. In marked contrast, some of the spine-tailed swifts have extremely short and blunt tails, so that they look like aerial torpedoes. It would be interesting to know how these differences in structure are related to differences in performance and manner of flight.

Much about the flight of birds remains to be discovered. While the broad principles involved in the use of wings have been set forth, few comparisons have yet been made between different types of wings and different types of flight in relation to the needs of the species concerned. The swift has sometimes been acclaimed as 'the best' of all flying birds, but this is to judge it by only one criterion, that of sustained speed in level flight. The jay, with its rounded wings and weak cumbersome flight in the open, is a master at turning through thick coverts without hitting the branches, its broad wings giving great control. The pheasant, which crashes through the twigs as it flies, is a master at rising in a steep and rapid climb from the ground. The eagle soars slowly in an upcurrent, the albatross sails swiftly on a gale, the hummingbird hovers with precision. These are the extremists and no one of them is 'best', for each is adapted to its own way of life. Further, there are the many more ordinary birds which can fly moderately fast, turn and check with moderate precision, hover rather clumsily, rise fairly quickly from the ground and, though relying on flapping, can make some use of upcurrents. We think of them as more ordinary because they are more numerous than the extremists, but their general-purpose wing, a compromise for several needs, is as well adapted for their way of life as are the specialised wings which attract a more obvious admiration.

Chapter 11

SWIFTS AT NIGHT

One evening we watched an open-air performance of *Cymbeline* in the quadrangle of All Souls. As it grew dark, Iachimo leant against a pillar, gazed at the sky and declaimed, 'Swift, swift, you dragons of the night', and at that moment several swifts flew screaming to their holes to roost. Because they need a clear drop in front to help them in taking off, swifts usually spend the night in holes in walls or cliffs; actually, they have more varied and more remarkable roosting habits than any other kind of bird.

The breeding pairs at the tower normally sleep in their boxes. They rest in a horizontal position, one on the nest and the other beside it. On cold nights they fluff out their feathers, appearing almost prickly, while one sometimes climbs on the back of the other, and thus they help to keep each other warm. There are published records of swifts roosting vertically, both in captivity and on a wall beside a nest, where the bird clung with its claws and pressed downward with its tail. The boxes in the tower are not deep enough to allow the birds to cling vertically, so we cannot say whether they prefer this to a horizontal posture.

Migrating swifts have a harder problem, as they may not know of suitable holes near where they happen to be at dusk. At times, they take their clue from residents. Thus on several occasions in spring and once in autumn, three adults have roosted together in a box at the tower, but only for one night at a time, the stranger presumably moving on next day. Although the owning pair vigorously expel strange swifts from their box during the daytime, they have at times allowed a third bird to enter after the normal time of retinue. One stranger was driven out of an occupied box shortly before most of the pairs came in for the night, but twenty minutes later it entered another box and snuggled in between the owning pair, and the three settled down amicably together to sleep.

In spring a passing swift has also come to rest at dusk clinging to the outside of the tower beside one of the holes, and in the autumn one entered an unoccupied box, when it acted very nervously. Other observers have seen single migrant swifts spending the night flat against a wall under the eaves, or on a window ledge. Another entered an upstairs window and roosted vertically in the folds of a curtain and yet another came in through a window and squeezed itself under the pillow of an occupied bed in a school dormitory. After a big passage of swifts along the Yorkshire coast on 20 August 1879, many spent the night on window ledges in Redcar. Likewise on 17 May 1955, in bitterly cold and wet weather ending with sleet, many swifts roosted on the outside wall of a school in Monmouth, and some came in at the top-floor windows and stayed inside all night, waking after some twelve to sixteen hours of apparently unbroken sleep. Similarly, on three nights in August 1954, swifts roosted on window recesses of the lighthouse at Dungeness, Kent. In each case the sky was overcast, the wind north-westerly and swifts had been moving along the coast during the day. On the nights concerned, 23, 5 and 14 birds were later caught at the light, and all proved to be young hatched that summer. One might expect young birds to migrate less strongly than adults. There have been similar records at Öland in Sweden after a big diurnal passage of swifts southwards against the wind, the birds coming to roost clinging to the lighthouse wall. Once the habit became known, ornithologists have placed nets against the lighthouse and these are pulled up after dark in order to ring any swifts that are roosting there. Nearly all have proved to be young birds.

A more remarkable roosting habit on migration was recorded at Scarborough in Yorkshire on 2 September 1897. A solitary swift came to roost clinging vertically to a small ash branch twenty feet above the ground, where it swung to and fro in the wind but evidently went to sleep for the night. Similarly on 12 September 1930, on an island in Hickling Broad, Norfolk, a single swift came to rest clinging to a tiny willow twig, and the observer was able to check that it stayed there all night. It arrived at 7.20 p.m. and left again between 7.15 and 7.45 a.m. on the following day. Few British swifts stay with us as late as September, and both these birds were probably migrants passing through from Scandinavia. It is therefore interesting that there are three other records of swifts roosting in the same way, one from Sweden and two from Finland. A photograph of one of these birds shows it hanging almost vertically, indeed partly upside down, beneath a twig. There has now been one further record of this habit, on 26 July 1953 in a swaying sycamore tree at Cromer in Norfolk.

More remarkable is the way in which passing swifts may roost in cold weather. On 25 and 26 June 1835, in tempestuous and unusually cold weather, huge numbers appeared at Dover in Kent, also at Walton-on-the-Naze and Harwich in Essex, and at all three places the birds were seen hanging in masses of up to fifty or sixty together on walls 'like swarms of bees'. The same was recorded at Deal in Kent on 8 July 1856, when after drizzle there was a very cold evening. Some of the clusters were two feet across, and in this and the earlier instance at Dover, many were knocked down and killed by children. Swifts were again seen clumping together at Dover on 1 September 1876, also at Ringwood in Hampshire in 1903 and on the Suffolk coast on 21 July 1930. This completes the published British records of the habit. Except for the undated Hampshire record, all have been from the east coast in unusually cold weather. As discussed later, swifts often travel for long distances in bad weather, probably coming from the continent to England, and it is likely that all the birds in question were strangers to the districts in which they spent the night.

On the Continent, much bigger clumps have sometimes been seen, notably during the extremely cold and wet summer of 1948. Thus on 7 July in that year, at Konstanz in Germany, after three days of unbroken rain and cold, a cluster of swifts included two hundred birds and another was more than ten yards long and one to two yards broad. Six days later in Basle, Switzerland, an observer saw on the front of a house six clusters, each containing between one and two hundred swifts. They looked like bunches of grapes. In May 1954, a cluster was seen to form when swifts were migrating north against a strong headwind in the Camargue, near the mouth of the Rhône. In the afternoon a swift came through a window into a room. It was caught and put outside. Soon afterwards the same or another bird entered in the same way, so the window was closed. After this a swift came to rest on the outside wall of the house against the corner of the window, and gradually others collected there until there were twenty, each touching its neighbours, and thus they spent the night.

The parallel with bees is apt. Bees clump together in their hives in winter in order to conserve their heat, and it is evident that swifts do so for the same reason. The habit has been seen only in cold weather, and the birds usually assemble earlier in the day and leave much later in the morning than from their normal roosts. Despite the extra warmth thus provided, swifts have often been found dead below the clusters by next morning. For instance, of the two hundred mentioned earlier which clustered at Konstanz, twenty were dead by the following day. These were perhaps the unlucky outside birds, and various observers have noticed how each individual tries to avoid the outside position by crawling further into the mass. During the cold summer of 1948, a small group of alpine swifts was also seen forming into a clump for the night in Switzerland.

To cling to the tip of a swaying branch, or to come together in clusters, is strange enough, but the swift has been claimed to have a yet more remarkable way of spending the night, which has been the subject of even greater argument than whether it mates on the wing. On warm and still summer evenings, both in the city of Oxford and in the surrounding villages, we have often seen swifts circling and screaming fairly high in the air; as the light wanes, they bunch more tightly together, fly with rapid beats, almost quivering the wings, scream more shrilly and rise higher, then the breeding adults break away and come down to enter their holes singly in the usual way, but still there is a screaming party high up, which gradually rises higher until finally the birds vanish from sight.

What happens to them? Various earlier observers claimed that they came down again to their holes soon after dark. This is almost certainly wrong. Swifts roosting on their nests often scream and may even scramble about after dark, and this noise has probably been mistaken by the observers for the return of swifts to their holes after dark. The question has been studied particularly by the Swiss observer Weitnauer. He attached an automatic recording device to his nesting boxes and has never yet registered the entrance or departure of a swift during the hours of darkness, even on those evenings when night ascents have occurred. He further pointed out that a swift finds it exceedingly hard to enter its hole in fading light, which we have also seen, and the other available evidence suggests that a swift cannot see at all well in the dark.

What then happens to the birds last seen, still rising, at dusk? The next step, clearly, was to watch in the early morning, and Weitnauer as well as two other observers had now seen swifts descending from a height down to the colony in the early morning, which strongly suggests that they have stayed up in the air all night. To carry the matter further, Weitnauer flew in an aircraft looking for swifts at dusk, and found several parties high in the air, both later in the evening than the time at which his breeding adults normally retire to their holes, and earlier in the morning than the time at which they normally emerge. These observations do not finally settle the matter, since the birds might conceivably return to the ground (though not to their holes) when it is too dark for them to be seen from an aircraft.

Fortunately, the matter can be carried further. Two reliable British observers told me that, in Cyprus and French North Africa respectively, they on a number of occasions in summer heard swifts screaming above them in the air long after dark, under circumstances that made it certain that the birds must have been in the air, not on or in high buildings. Recently, also, two Swiss observers have seen birds that were almost certainly swifts crossing the field of a telescope pointed at the moon on summer nights, though it is hard to be sure of the identification under such conditions. The birds were estimated to be in one instance at 7,000 feet and in the other at between 3,000 and 7,000 feet. Finally, there is one definite observation by a French airman in the 1914–18 war, an account which until recently was by most people dismissed as absurd. One night he was on a special operation on the Vosges front which involved climbing to 14,500 feet above the French lines and then gliding down with engines shut off over the enemy lines. 'As we came to about 10,000 feet, gliding in close spirals with a light wind against us, and with a full moon, we suddenly found ourselves among a strange flight of birds which seemed to be motionless, or at least showed no noticeable reaction. They were widely scattered and only a few yards below the aircraft, showing up against a white sea of cloud underneath. None was visible above us. We were soon in the middle of the flock, in two instances birds were caught and on the following day I found one of them in the machine. It was an adult male swift.' (Author's translation from French original.)

It is on the wing that she takes her repose,
Suspended and poised in the regions of air;
'Tis not in our fields that her sustenance grows,
It is winged like herself – 'tis ethereal fare.

1: Adult swift incubating. (Photo Manuel Hinge)

2: Adult near the entrance hole to its nest, showing the eye set deep in the head for protection in fast flight. (Photo Manuel Hinge)

3: Adult flying showing the pale throat patch. (Photo Steve Blain)

4: Adult rapidly changing direction in flight showing spread wings and tail raised and fully splayed. (Photo Steve Blain)

5: Oxford University Museum. The west and south sides of the tower are visible, each with ten cowls. There are two swift boxes under each cowl. Note that the carvings around the windows were never completed. (Photo Colin Wilkinson)

6: A screaming party. (Photo Colin Wilkinson)

7: Adult bringing a feather as nest material. (Photo Manuel Hinge)

8: Adult with throat distended as full of a bolus of small insects for the newly hatched young. (Photo Manuel Hinge)

9: Adult feeding newly-hatched young. (Photo Manuel Hinge)

10: Young about seven days old showing the feathers starting to erupt and unopened eyes. (Photo Manuel Hinge)

11: Young about fifteen days old with eyes starting to open and feathers growing. (Photo Manuel Hinge)

12: Two young about fifteen days old with feathers growing and a discarded egg beside the nest. (Photo Manuel Hinge)

13: Well-grown young showing extensive white around the beak. Adult on right.
(Photo Manuel Hinge)

14: Well-grown young showing white edges of head and wing feathers.
(Photo Manuel Hinge)

15: Adult approaching nest. Photograph taken from the nest entrance.
(Photo Manuel Hinge)

16: Adult about to enter nest with wings and tail spread to brake and feet forward.
(Photo Manuel Hinge)

17: Young swift showing very short beak but large gape.
(Photo Manuel Hinge)

18: Flat fly, *Crataerina pallida* on young swift. (Photo Manuel Hinge)

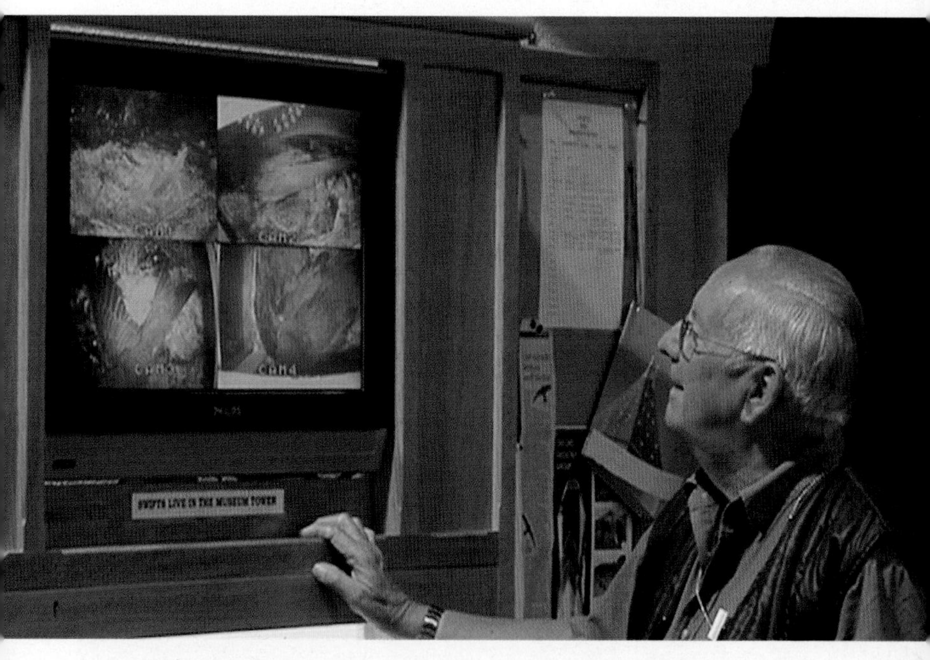

19: Roy Overall, long-term keeper of the swifts, by the video link in the main court of the University Museum. (Photo Manuel Hinge)

Cowper's verse, a century and a half old, might seem prescient of the modern discovery, but it was written of the swallow, not the swift. I am convinced by Weitnauer's evidence that swifts do regularly spend the night on the wing in fine weather. He has often inspected his nesting boxes during the night after an evening ascent has occurred, finding the breeding swifts present on their eggs or young but the non-breeding yearlings absent. Clearly it is the birds without family ties that take part in the night ascents, and only rarely has a breeding adult gone with them. In May, however, before breeding has started, we have not infrequently found one and sometimes both members of a pair away from their box in the tower at night. Some individuals have been much more prone to do this than others.

While watching nightjars during a short visit to the Suffolk coast in 1952, we on 2 July took an evening off to see an avocet; and chanced on something even more pleasing. At dusk, ten swifts flew high over us and out to sea. We supposed that they might be birds returning to the Continent after one of the weather movements described in Chapter 13, and did not follow up the observation. But two years later we again relaxed on our last evening, this time to see eight spoonbills, and again saw something more rewarding. As the light failed, thirty-three swifts stopped feeding over the marshes and headed out to sea, being lost to sight at the extreme limit of range of the field-glasses.

This second incident suggested that the habit might be regular, so we kept special watch for it during a month of the summer of 1955, and found that swifts not infrequently set off seawards at dusk, and in every instance they continued to fly away from the land until lost to sight. On a number of evenings, swifts were seen hunting for food above a belt of trees about half a mile inland bordering the coastal marshes. Shortly before it grew dark most of these birds flew off inland, presumably to their nests, but others sometimes stayed for about ten minutes longer, then turned and headed for the sea-wall, screamed once or twice as they crossed it, and flew on silently out to sea. They usually kept one or two hundred feet above the water, flying either directly or in small circles away from the land, and making no attempt to return. The number leaving together in this way varied from two to about a hundred.

On one evening in heavy rain, there was a more dramatic incident. Swifts were pouring south along the coast in a weather movement, at a rate of 1,500 in five minutes. As it grew dark the passage stopped, so abruptly that it seemed as though a referee had blown the whistle for 'time'. The swifts near to me, about 120 in all, circled several times over the beach, gaining a little height, and then set off out to sea through the driving rain. As no further swifts passed along the coast, similar parties were presumably leaving seawards at intervals all along it. On another evening there was a small passage down the coast, and again the movement stopped at dusk and those swifts near to me circled and went out to sea.

Night ascents from the breeding colonies take place only on fine and still evenings, the birds rising high in small circles, with rapid quivering wing-beats and intense screaming. In marked contrast, the dusk flights out to sea have taken place in all types of weather, on a calm and clear evening with a full moon, with a fresh wind and overcast sky, and in pouring rain. Further, the manner of flight has been normal, the birds did not rise much above a hundred feet, and were silent except for one or two screams at the moment of leaving the land. Clearly, the seaward flights have little in common with the night ascents inland, except that in both cases the birds presumably spend the night on the wing. These seaward flights at dusk were seen chiefly when there had been a weather movement, either immediately beforehand or earlier on the same day. It is reasonable, therefore, to suppose that the birds taking part were strangers and did not know of suitable roosting holes in the area. Several reports were described earlier in this chapter of swifts clumping together in cold weather on the east coast. Probably, they also were birds on weather movements, but since most weather movements do not end in this manner, swifts evidently prefer to roost on the wing when the weather is not too unfavourable. It seems strange that they should go out over the sea, but they are there free from projecting obstacles such as trees or hills, and at night there is a greater likelihood of rising currents over the sea than over the land.

Two points remain for discussion. First, can birds sleep on the wing? The swift with its long thin wings is adapted for flying at high speed. It has been suggested that it might stay up at night by gliding in thermals, but it has the wrong type of wing for slow-speed soaring. When it glides it must descend through the air rather rapidly, so that it could not stay aloft without fairly frequent flapping. All the same, there seems no reason why a swift should not get enough sleep in this way, by periodic flapping to gain height, then gliding and taking a 'cat-nap' until the next period of flapping is needed. We should not, perhaps, take too human a view of the need for long and sustained repose. For instance, in the continuous daylight of an Arctic summer there is always some noise at sea-bird colonies, so probably the birds get their sleep in short spells. The second question to ask is why it is necessary for swifts to sleep aloft. The birds seem safe in their holes, and to stay up all night must use up energy. Weitnauer has suggested the probable answer, that on migration and in their winter quarters swifts may not know of suitable roosting holes, and then the ability to repose in the air might be of great value. Night ascents have been reported from Africa, and C. W. Benson writes from Rhodesia that he cannot imagine where the swarms of swifts that appear with the flying ants could spend the night unless they stay in the air.

It is curious that no other species of swift has yet been reported ascending at night. One might have expected the habit to be found, at least, in some other members of the genus *Apus*. Perhaps it is needed only in the migratory species, since those which are permanent residents could roost in their nesting holes throughout the year. Since most swifts appear to be permanent residents, this might restrict the habit to a very few species.

Other swifts, so far as they have been studied, normally roost in situations similar to their nesting sites. The species of *Apus* usually roost on cliffs, though in India a group of house swifts was seen using old nests of baya weaver-birds during a rainy spell, abandoning them when the weather improved. The scissor-tailed swifts (*Panyptila*) use their own hanging nests for shelter from rain and also, presumably, for roosting at night. Among the black swifts (*Cypseloides*) of tropical America, cloud swifts have been seen by several observers dashing through waterfalls at dusk to roost in great numbers on the rocks behind, while chestnut-collared swifts have been found roosting in a compact mass of thirty to forty individuals on the vertical side of a rocky cliff. Both these species and two others have sometimes come to lighted windows on foggy and wet nights at William Beebe's laboratory at Rancho Grande in Venezuela. As is well known, migrating birds sometimes strike lighthouses on foggy nights, but while these swifts might have been migrating, is it possible, instead, that they had ascended for the night, like common swifts?

The American chimney swift, it will be remembered, nests in hollow trees or chimneys, and it roosts in similar situations, but with this difference, that outside the breeding season the roosting flocks may be of great size, and hence they need large trees or large chimneys to accommodate them. Fortunately, after Americans had felled the giant trees of the virgin forests they built factory chimneys, which provide similar roosting conditions for the chimney swifts on their migratory journeys. Audubon, more than a century ago, saw about nine thousand chimney swifts roosting in a single hollow sycamore tree. The tree was over sixty feet high, and five feet wide at forty feet above the ground. He cut a hole in the base of the tree, crawled through the excrement and shone a light, to see the birds clinging side by side in the interior. Rising before dawn, he placed his ear against the outside of the trunk and, after waiting twenty minutes, heard such a rushing noise that he sprang back thinking the tree was falling, but it was merely the chimney swifts leaving for the day.

A modern observer has counted over twelve thousand birds roosting in one chimney. The aerial evolutions of a flock round the chimney chosen for the night are more wonderful than those of our own roosting starlings in Trafalgar Square. The chimney swifts sweep round in circles and, as they pass over the chimney, some of them drop lower, spreading the tail and pressing it downwards as far as possible, at the same time raising the wings high above the back and fluttering them through a comparatively small arc, then dropping directly, or turning jerkily, or twirling gracefully into the hole. On leaving, the birds have to fly upwards and they do this with vigorous flaps holding the body almost vertically, 'as if crawling up invisible wires'. At the present time, thousands of migrant chimney swifts are ringed (banded) each year by placing a catching device over the chimneys where the birds roost.

Chimney swifts have once been found roosting in dense elderberries. Once also, a hundred or more were seen clumping together on the bark of an oak tree some twenty-five feet above the ground, where they formed into a cluster five feet high and seven to eight inches wide. Here they remained from the late afternoon of one day, through the night, until well into the following morning. This recalls the similar habit of the common swift.

No other kind of bird is known to spend the night on the wing like our swift. But clumping in cold weather is also found in bee-eaters, the dusky wood-swallow of Australia, and in European swallows and martins. Again, long-tailed tits spent cold nights bunched tightly together in line on a branch and wrens pile into convenient cavities. As many as seventeen wrens have been seen to enter one small nest-box for the night. Aerial manoeuvres before roosting, like those of the chimney swift, also find parallels in other birds which roost together in large numbers, such as starlings. It has been suggested that they may serve to distract birds of prey, which might otherwise take great toll from the dense flocks. But these and the many other aspects of roosting and sleeping in birds require much further study.

Postscript: Since this book went to press, I have confirmed the suggestion on p.128 that swifts on passage might sleep on the wing, by observing mass ascents at dusk from Tring reservoirs, both in May (true migrants) and June (during weather movements). Nearly all those feeding over the water rose up together as it grew dusk, with a few screams, and after reaching a height of one or two hundred feet, circled and drifted away together.

Chapter 12

WINTER SLEEP

There has been one other grand controversy concerning the swift, and though it has now died down it should possibly be revived. The annual disappearance of birds in autumn and their return in spring is not easily explained by anyone whose knowledge is restricted to one place. Bird migration was known in ancient times and is mentioned in the first serious work of ornithology, that of Aristotle, but Aristotle supposed that certain other kinds of birds spent the winter in hibernation, including the swallow, kite, stork, turtle dove, thrush and lark, while Pliny added the cuckoo. On the knowledge available at the time this view was not unreasonable, especially since bats pass the winter in hibernation and bats were for long classified with birds.

By the middle of the eighteenth century, bird migration was much more widely recognised and in Europe only the swallow tribe and swifts were still credited with hibernation by serious naturalists. Gilbert White, it will be recalled, spent much time and thought on the problem and even got labourers to dig in a likely place for wintering swallows, though without success. He also knew of a clergyman who in the early spring saw two or three swifts that had been found torpid but alive in the brickwork of a church tower. Further, in the 1776 edition of Thomas Pennant's *British Zoology*, which long remained a standard work, it was recorded that in February 1766, a pair of swifts was found torpid but alive in Longnor Chapel, Shropshire, and Pennant suggested that this was how the species might spend the winter. Gilbert White further noted that the black feathers of the swift become gradually more bleached during the summer, yet the birds reappear next May in fresh plumage. Could they not, he therefore asked, retire to some safe place for the winter to moult? There is also a later record of a swift found torpid in midwinter, in an old chimney in Bolton Hall, Yorkshire. On the Continent, too, there were several hearsay accounts of swifts hibernating.

Actually it was another eighteenth-century naturalist who, like Gilbert White, chose to spend his life in a village, who gave the correct answer for the swift. This remarkable man was also the first to describe how the nestling cuckoo ejects the eggs or young of its foster-parent and the second person to mark any species of wild bird for scientific purposes, while he did such good work on the natural history specimens brought back by Captain Cook's first expedition that he was invited to go on the second voyage. This he refused, as he felt called to his medical work at home, and as a result, he made a far greater discovery than any likely to have come from a voyage with Captain Cook. For Edward Jenner, living quietly at Berkeley in Gloucestershire, invented vaccination, and the resulting work and correspondence took up so much of his time that even his important paper on the migration of birds was not published until his nephew submitted it, after his death, to the Royal Society. In this essay, Jenner described how he marked several swifts by cutting out their toes and how they returned to the same dwelling in several later years. He, contrary to Gilbert White, was convinced that they migrated and his soundest argument was that, if they emerged in spring after hibernation, they should be weak and thin, whereas in fact the swifts that he examined just after their return were fat and in excellent condition, a point that Gilbert White overlooked.

At the present day, we know that common swifts appear in Africa during our winter, and there have been several recoveries in tropical Africa of individuals ringed as young in Europe. Hence it is clear that they migrate. In cold spring weather, as mentioned in the last chapter, swifts have occasionally been found half-torpid, but there is no reason to think that they can survive in this condition for more than, at most, a few days. The few published accounts of hibernating swifts were all at second or third hand, and it seems not impossible that the original observer had forgotten the month in which he found the birds and, prompted by a too eager questioner, thought that it must have been during the winter when in fact it was in the spring. Despite the great increase in bird-watchers, there have been no recent records of swifts found torpid in winter, so it may reasonably be concluded that swifts do not hibernate in Britain.

Swallows were also thought to hibernate. The ancient writers, including Aristotle, supposed that they retired for this purpose to caves, cliffs or hollow trees, but a far more extraordinary story was spread abroad in the sixteenth century, which apparently started with Olaus Magnus, Bishop of Uppsala in Sweden. 'Touching their lurking places, Olaus Magnus maketh a farre stranger report. For he saith, that in the North parts of the world, as Summer weareth out, they clap mouth to mouth, wing to wing, and legge to legge, and so after a sweete singing, fall downe into certaine great lakes or pooles amongst the Canes, from whence at the next Spring, they receive a new resurrection; and hee addeth for proofe thereof, that the Fishermen, who make holes in the ice, to dip up such fish with their nets, as resort thither for breathing, doe sometimes light on these Swallowse, congealed in clods of a slymie substance, and that carrying them home to their Stoves, the warmth restoreth them to life and flight!' This story was strongly supported up to the middle of the eighteenth century, many country people giving affidavits that they had found swallows in this state. It was recorded in print by, among others, the great Linnaeus, though when he was questioned about it by Peter Collinson, a Fellow of the Royal Society, Linnaeus ignored him and evaded the experimental proofs suggested to him. Experiments were in fact performed by the Abbé Spallanzani, but with a negative result. He put swallows in wicker cases under snow and found that they did not go torpid but, after a time, died. The idea of hibernation under water possibly arose through the swallows' habit in cool weather of collecting over water to feed. It has also been suggested that the birds reported to Olaus Magnus as plunging under the water in the northern countries were dippers, which in colour are not unlike house martins.

For the swallow, as for the swift, bird-ringing settled the matter, though one can scarcely believe the first reported recovery of a marked swallow in its winter home. Caesarius von Heisterbach, prior of the Cistercian Monastery at Koenigswinter in Germany, in his book *Dialogus magnus visionum et miraculorum* edited between AD 1219 and 1223, stated that a man took an adult swallow from its nest and fixed to its foot a parchment with the note 'Oh, swallow, where do you live in winter?' Next spring the swallow returned, bearing another note which read 'In Asia, in home of Petrus'. More relevant is the experiment in the middle of the eighteenth century by Johann Frisch, who put red threads round the legs of swallows. If, he argued, they went beneath the water for the winter, the red dye should come out of the threads. In fact, when the swallows returned next year the threads were still red, so he concluded that the birds did not submerge. Frisch may fairly be claimed as the first person to mark wild birds for scientific study, with Jenner on the swift a close second. In recent years, of course, there have been numerous recoveries in Africa of swallows ringed in Europe in summer. Swallows and martins, like swifts, may become temporarily torpid in cold weather, and this was perhaps one source of the unfounded but formerly widespread belief that they hibernated.

Similar accounts have come from America, for both swallows and swifts. On 19 December 1879, fifty-five days after he had seen the last chimney swifts on the wing, C. C. Abbott, a reputable New York ornithologist, found seven of these birds 'snugly stowed away in a (disused) stove-pipe'. When placed on the floor they soon recovered and after a feeble flight about the room they left through a window and were seen no more. This occurred during an exceptionally mild winter. Another chimney swift emerged from a chimney and entered a room in Ottawa in February 1883, the bird being caught and remaining alive for several days. The observer was a zoologist and geologist, and there is every reason to suppose his report accurate. The American ornithologist W. L. McAtee, who summarised these and a few similar records, has himself handled two torpid chimney swifts, though not as late in the year as these others. As he pointed out, the records do no more than provide convincing evidence for temporary torpidity. One wonders whether, in the instance of 1879, the birds came out periodically and found food in the mild winter weather, but it is astonishing that another chimney swift should have survived until February so far north as Ottawa. Despite the great increase in American bird-watchers, there have been no more such records in the last forty years. It is also clear that almost all, if not quite all, North American chimney swifts migrate south for the winter. Hence it is doubtful that they ever hibernate and, if they do so, it must be extremely rare. But the possibility cannot be ruled out.

There is a much more striking record for the white-throated swift, which is resident throughout the year in parts of California (though a full migrant further north in its breeding range). W. C. Hanna reported that at Slover Mountain in San Bernardino County, California, 'during the extremely cold wave of early January, 1913, eight, to me perfectly healthy, (white-throated) swifts were taken out of a crevice where they, with many others, seemed to be roosting in a dazed or numbed state'. They were kept for six hours in a room, turned loose and then flew away, still rather dazed. Since white-throated swifts apparently stay all the year round in this area but are out of view for days at a time in the coldest weather, temporary hibernation is perhaps a regular feature of their life there. Unfortunately the species roosts, as it nests, in almost inaccessible crevices on cliffs, so that this interesting possibility has not as yet been studied further. It suggests that the torpidity of swifts might be an adaptation not merely for surviving the rare cold day in spring or autumn, but for enabling them to pass the winter further north than would otherwise be possible, in a climate where brief cold spells are the rule in winter. If this holds for the white-throated swift, could it perhaps hold for other kinds of swifts at the appropriate latitude?

From America also, in former years, came stories of hibernating hummingbirds, but these doubtless arose from their habit of temporary torpidity, now well established. There were also circumstantial accounts, though not by ornithologists, that Carolina parakeets clustered in large numbers inside the hollow trunks of trees in cold weather, but the species is now extinct, so the point cannot be checked.

When, in 1947, W. L. McAtee reviewed the published accounts of hibernation in birds, he concluded that most records were entirely false, though there was reliable evidence that various kinds of birds could survive in a torpid state for short periods and there were a few records, apparently to be trusted, suggesting that such torpidity might sometimes be more prolonged. There was no evidence for true hibernation in any bird, though some of the records were more suggestive than most naturalists were at that time ready to admit. The review was prescient, for it had scarcely appeared when a true case of hibernation was recorded in the poor-will, a nightjar which gets its name from its call-note. The bird was found torpid in a deep crevice on a cliff in the Colorado Desert in California in December 1946. The place was revisited in 1947 on 26 November, when the bird was found there again, and it was later visited almost every weekend until 22 February 1948, when it had become active. Hence it hibernated for at least three months, during the season when the night-flying insects on which it feeds are scarcest. When handled by the observer it took no notice, so that its temperature and weight could be measured without disturbing it. Whereas most birds have a temperature around 105°F, successive readings for this bird gave figures around 66°F. Hence there was a great reduction in its bodily activity, and this of course is the value of hibernation, in allowing survival without food. Song-birds of similar size to a poor-will can normally survive without food for at most a few days, during which they lose up to half their weight. This poor-will survived at least three months, and in the six weeks for which it was weighed it lost only one gram, its weight decreasing from 45 grams on 4 January to 44.6 grams on 14 February. Unluckily, it was no longer there in the following winter. A pleasing footnote to the discovery is that it is not new, for the Hopi Indians call the poor-will 'holchko' which, translated, means 'the sleeping one'. How the news would have delighted Gilbert White.

Now that one hibernating species has been found, the possibility of others should be brought back to mind. The habit seems more likely to be evolved not in the far north but in countries which, like California, enjoy a Mediterranean type of climate with a relatively mild winter. It will therefore be recalled that the original accounts of hibernating birds came from the Mediterranean region. Nevertheless, that any European bird should hibernate seems extremely doubtful, though it would be pleasing to find one that did.

Chapter 13

WEATHER MOVEMENTS

On a hot July day we were walking along the top of the Berkshire Downs when huge thunderclouds came up from the west. Shortly before the rain broke over us, a party of swifts flew low past us away to the east. The black birds under the black clouds seemed fitting heralds of the storm, though they were of course avoiding it, not welcoming it. Soon afterwards we were soaked through and envied the birds their power to escape.

Swifts avoid rain whenever possible. From the hills above Oxford in the wet summer of 1954 we often saw parties of up to six hundred swifts drifting in irregular circles away from approaching thunderstorms. This habit is so well known that it has passed into folk-language, and in several parts of Europe the swift is locally called a 'rain-swallow' or 'thunder-swallow'. Again, when we were watching outside the village nests in the early years of this study, we were sometimes drenched by a passing shower but, a few minutes after it had passed, would see parent swifts returning to their nests with dry plumage, having flown round the area of rain. If we are sitting in the tower and hear sudden rain on the slates outside, several swifts usually come in to their nests for shelter almost at once and stay until the rain stops.

> The swift-winged swallow feeding as it flies
> With the fleet martlet thrilling through the skies
> As at their pastime sportively they were
> Feeling th' unusual moisture of the air
> Their feathers flag, into the ark they come,
> As to some rock or building, their own home.

Samuel Drayton in *Noah's Flood* (1612)
(Martlet is the old word for a swift.)

It is an advantage derived from their long narrow wings that swifts can fly round storms, though it is a disadvantage of their long narrow wings that they can feed only in the air and hence are deprived of most of their food supply during rain. As pointed out in Chapter 9, most insects leave the air in bad weather, so it is interesting to know how other British birds which prey mainly on adult insects obtain their food on rainy days. In the first place, birds of this type are much commoner in dry than wet climates. The bird-watcher in North Africa is immediately struck by the abundance of shrikes, bee-eaters, rollers, swifts and other predators of winged insects as compared with England. Nevertheless, some of the birds which prey on air-borne insects breed successfully in Britain and each adopts a different technique for wet weather.

No other species flies round storms like the swift. In rain, swallows hunt close to trees, buildings, water or the ground, where small insects are still to be found. As already pointed out, the swallow's greater skill in manoeuvre allows it to fly safely where a swift could not come. Spotted flycatchers, which in fine weather catch insects in the air by flying out from a high perch, in rain adopt the same method under trees, where some insects are still on the wing under this natural umbrella. Red-backed shrikes are more ingenious, catching more insects than they need on sunny days and impaling them on thorns in a larder, from which they take them in bad weather. It was formerly supposed that these larders were useless because the shrikes did not remove the insects from them. In fact they often remove them in wet weather; the error perhaps arose because ornithologists watch mainly in fine weather.

The direct avoidance of showers and thunderstorms is not the end of the story so far as the swift is concerned. At midday on

9 June 1953, when we were sheltering from a shower on the beach near Southwold in Suffolk, we were suddenly aware of swifts coming in low over the sea. In the next ten minutes two or three hundred arrived. They were visible out to sea to the furthest limit at which field-glasses could detect them, flying low over the water against the westerly wind and proceeding straight on inland in the same direction. If they had kept to this westerly course the whole time, they must have crossed a hundred miles of sea from Holland. July 9 is, of course, too early for the start of the swift's autumn migration and in any case the birds were flying west not south. So what were they doing?

This is just one example of a type of summer movement of which there were only a few reports before 1939 but many in the last few years since attention has been drawn to them. The earliest published record is for the year 1873, when large numbers of swifts passed west along the sea-front at Brighton on 30 June, while on the next day unusually large numbers were seen moving in the same direction over London, which suggests that the movement was extensive. Between 1929 and 1939, the bird-watchers at Tring reservoirs in Hertfordshire often reported swifts drifting west in the latter part of June and early July, usually in unsettled weather, and wondered what they were doing. Most of the other British records have been on the east coast, between Spurn Point in Yorkshire and Dungeness in Kent; in four instances, three at Cley in Norfolk and one at Gibraltar Point at the mouth of the Wash, five thousand birds have passed in a day, though from five hundred to a thousand has been more usual.

Although these movements occur at the height of the breeding season, they were formerly attributed to migration, a late spring migration if the birds were travelling more or less northwards and an early autumn departure if they were moving south or southwest. This even led to the absurdity of spring migration on one day followed by autumn migration on the next day, at the same place.

Similar movements in summer occur in the lands round the Baltic, where they at times involve much greater numbers than in England. The biggest passage so far recorded was on 4 July 1947, on which 27,000 swifts travelled south past the southern tip of the Swedish island of Öland, 14,500 having passed in the same direction on the previous day.

The Finnish zoologist J. Koskimies analysed some of these spectacular movements in relation to the weather maps, and found that nearly all occurred in the south-eastern sector of an approaching depression, which is the sector in which most rain falls. Often the birds were on the move before the rain reached the place where they were seen, and normally they flew more or less against the wind, heading between southeast and southwest. Koskimies pointed out that, if the birds continue flying into the wind, they will pass round the southern side of the depression, in the warm sector, and thus will get out of the rainy weather more quickly than if they stayed where they were and waited for the rain to pass. These 'weather movements' may therefore be regarded as a more elaborate way of avoiding rain than the direct flights from local showers already described.

Later observations in Sweden have given general support for Koskimies' theory. At the migration station on Öland, in the three summer months of June, July and August in the years 1947–9, there were forty-seven movements of swifts involving over a thousand birds, and six of them involved over ten thousand birds. It was found that nearly all of these movements occurred during the passage of a depression, the others (only four) being comparatively small movements in August, probably true migration. Most of the movements took place on the day after rain had fallen on the Swedish mainland and a few of them after it had fallen in southern Finland. The birds travelled against the wind, except that, like true migrants, they tended to follow a coastline where it ran more or less in the direction of their course. On Öland this means that they normally flew southwards, but there was one exception, on 25 July 1948, when a depression centred over Poland moved northwest instead of the usual northeast, resulting in a strong north-westerly wind on Öland, and many swifts passed north. As usual, they were flying against the wind and away from the depression. This last instance, and evidently some of the others, were exceptions to Koskimies' view that the movement normally occurs in the south-eastern sector of a depression.

After the passage of a depression, one would expect the swifts to return to the place from which they started, which will usually involve a northward flight in fine weather. There are a number of records to support this idea, but far fewer than of southward movements against the wind, which Koskimies reasonably suggested is because in fine weather swifts fly high. Probably also, there is not the same tendency to concentrate along coastlines, hence the parties will tend to be much smaller and less conspicuous.

In England, as already mentioned, the summer movements of swifts are on a much smaller scale than in Scandinavia, and attempts at analysis by previous workers did not give support to Koskimies' theory. To test the matter further, I studied the weather maps for all the days on which large summer movements of swifts were recorded in Britain during the years 1946–53, and found that, except in one point, nearly all of them fitted with the pattern suggested by Koskimies.

During the summers mentioned there were, in all, eighteen movements involving at least a thousand swifts in a day. As many as two-thirds of them were observed either southwards in Suffolk or westwards in north Norfolk, and four others took place on the Kent or Sussex coasts. Hence most have occurred in that part of England nearest to the Continent. All save one of them occurred when the wind was between south and west and the direction of flight of the birds was between south and west. In about half of them the birds were flying directly against the wind, and in all save one of the rest they flew within 45 degrees of this direction. Moreover in the one instance of a northerly flight, the wind was from the north at the time. Hence the movements normally take place against the wind, the only exception occurring two days after the northerly passage just mentioned, when many swifts were seen flying south with a cross-wind from the west, but they were perhaps birds returning from where they had come two days before.

Only three of the movements occurred in exactly the circumstances postulated by Koskimies, that is, southwards in the south-eastern sector of an approaching depression. An instance of this is shown in Fig. 16, with a depression centred over northern Ireland and swifts flying south against the wind in the mouth of the Wash and along the Suffolk coast. But ten other movements took place to the south of a depression centred over Scotland or the North Sea, and another when rain was falling in Scotland. Hence the general connection with depressions is clear, though the movements have occurred much more often to the south than to the southeast of their centre. The one instance of a northerly movement is of special interest, as it took place away from a depression to the south. From dawn onwards on 23 June 1951, large numbers of swifts were seen flying northwards, arriving off the sea, on the south coast of England at Poole Harbour in Dorset and near Eastbourne and Hastings in Sussex (and presumably in between). The weather map at midnight previous to this passage, in Fig. 17, shows that there was at the time a complex thundery depression centred over northern France and Belgium. This instance recalls the exceptional northward movement on Öland when a depression was centred to the south, over Poland. Finally, three of the movements were more puzzling, in that they were not associated with depressions, one occurring in unsettled weather with scattered showers, the others in rather settled weather with high pressure over England. In large part, however, Koskimies'

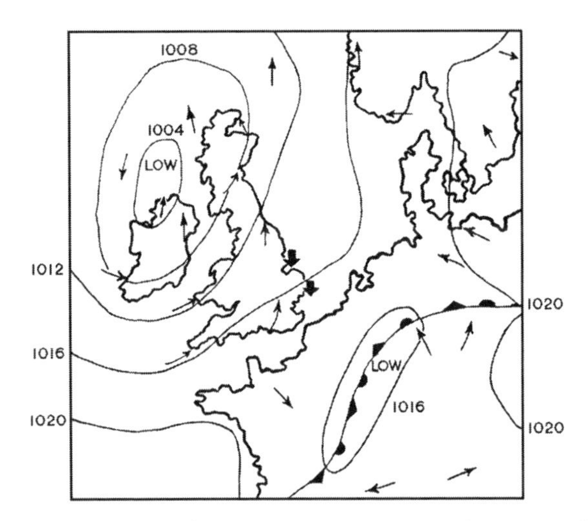

Fig 16: Weather map for 6 a.m. on 17 June 1951. Thin arrows show direction of wind; thick arrows of moving swifts

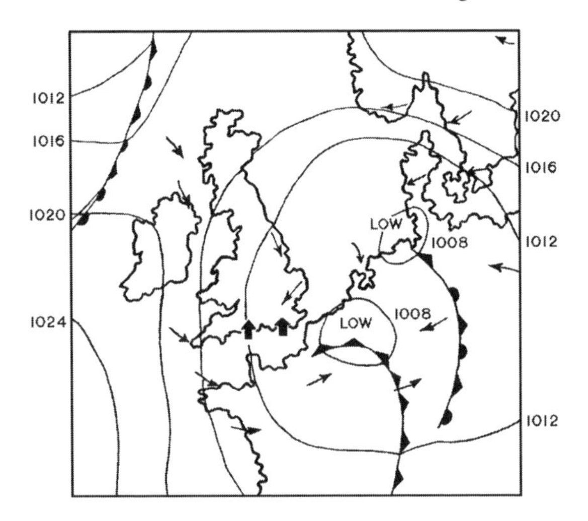

Fig 17: Weather map for midnight on 22 June 1951. Thin arrows show direction of wind; thick arrows of moving swifts (at 6 a.m.)

(Reproduced from 'Daily Weather Reports' by permission of HM Stationery Office and the Meterological Office)

theory that the swifts are flying away from depressions has been substantiated.

Just after completing this analysis, we spent four weeks of the summer of 1955 on the Suffolk coast in the hope of seeing weather movements in progress. For nearly the whole of this time pressure was high, fine weather prevailed, and no swifts passed, but on 2 July a depression was reported west of Ireland, it moved in east, deepened, and reached the Lake District by 6 a.m. on 3 July. In the period when Suffolk was in the south-eastern sector, no movements occurred, but several thousand swifts travelled south along the coast on the evening of 3 July and a few hundred did so on the following morning. Further, on 5 July, after the depression had passed and the weather was fine with a northerly wind, we saw small parties of swifts, in all eighty-five birds in three hours, moving northward, presumably on their return flight, which has rarely been seen before.

Since these movements happen during the breeding season, Koskimies postulated that it is mainly yearling swifts that take part, as they are without family ties. Supporting this idea, the Swiss worker Weitnauer found that during the summer the breeding adults did not normally leave their colony, but the non-breeding yearlings often vanished for several days and nights together. As to what sets the birds moving, a direct response to falling rain seems ruled out, since Koskimies reported that swifts may be on the move in an area ahead of the rain-belt. He suggested that they might respond to a reduction in air-borne insects, but I think it more likely that they react, as we do ourselves, to simple signs of impending bad weather, such as an overcast sky with low cloud and a moderately strong wind, though in swifts such a response is presumably inborn, not learnt. Whatever sets them moving, and it might be a combination of factors, the swifts then travel against the wind except for minor deflections to follow a coastline, and if they keep moving against the wind they will reach better weather.

It is understandable that these summer movements should involve fewer birds in England than in Scandinavia. In England the weather varies more rapidly and more irregularly than in Scandinavia and summer rain is not usually associated with great cold. Hence rain has a briefer and a less harmful effect on airborne insects than it does in Scandinavia or in central Europe. Indeed one meteorologist whom I consulted thought that in unsettled weather an English swift might do better to stay where it was and wait for the bad weather to pass, rather than to fly a long way, using up much energy, which might merely bring it into another area of rain. It is therefore interesting that the position and direction of the English movements suggest that many of the birds taking part may have crossed from the Continent. Thus swifts moving west into East Anglia and north into Sussex from the sea have presumably come from Holland and France respectively.

Little definite is known of the distances for which swifts travel on such movements. The arrivals from the east into Suffolk suggest that swifts may be prepared to cross a hundred miles of sea. On Öland, as already mentioned, movements have usually occurred with rain in central or southern Sweden, but sometimes with rain on the eastern side of the Baltic and even in southern Finland, three hundred miles away, and it may be supposed that the birds came from where the rain fell. More precise evidence is provided by the recovery of eight swifts caught and ringed on Öland during weather movements between 6 and 12 July 1947. These were found in subsequent summers, very likely on their breeding grounds, five of them in central Sweden, 150 to 250 miles to the north of where they were caught, and three of them in central Norway, 330 to 375 miles to the north of where they were caught. In addition, two ringed at Öland on 1 July 1950 have been recovered in a later summer 150 to 175 miles to the north, in central Sweden, while one ringed on 17 June 1949 was later found 275 miles to the south, near Lübeck in Germany. These records suggest that swifts may move up to at least 375 miles from their homes, so that a complete weather movement may well involve a journey of a thousand miles.

The most distant recovery of a swift in summer is one of our own Oxford birds, which was ringed as a nestling in 1951 and was killed on 17 July 1952 in the air-intake of a jet aircraft flying at the time over Jutland in Denmark, rather over 500 miles to the northeast. In this case it is not certain that the bird had taken part in a weather movement, especially as the weather just beforehand was not of the type which usually gives rise to such movements. But it was a yearling and it seems unlikely that it would wander so far north in the breeding season except on a weather movement.

This ends the story so far as it has been carried in the common swift, but there is one similar record for the alpine swift. This species has occurred in England not infrequently on migration, usually singly. On 15 July 1917, at the height of the breeding season, a flock of about a hundred was seen flying north-westwards against the wind in heavy rain at Kingsdown on the east coast of Kent. The weather records for this period, though incomplete owing to the war, suffice to show that a depression moved in from the Bay of Biscay eastwards across southern France on 13 July, reaching northern Italy on the following day. The nearest breeding places of the alpine swift are in the Pyrenees and in Switzerland, both of which must have experienced bad weather in this period, and the birds presumably came from one of them, involving a journey of at least 400 miles. Much cannot be argued from one incident, even though it involved many birds, but this has all the appearance of a weather movement of the type found by Koskimies for the common swift. In one respect it may have been abnormal, since a few alpine swifts stayed in the district for another three weeks, possibly because they were lost.

Similar movements have not been reported in the North American chimney swift, but the appearances of the black swift (*Cypseloides niger*) at Vancouver have recently been shown to be correlated with the approach of a depression from the Pacific, the birds, as usual, flying against the wind. The story for this species seems essentially similar to that found in the common swift. The large needle-tailed swift, which migrates from eastern Asia to Australia for the winter, is likewise said to be seen most commonly in its wintering grounds during periods of unsettled weather.

In the hot and dry regions of Africa, great gatherings of swifts are again associated with rain and thunderstorms, but for the opposite reason from that which holds in cool climates, since in central and southern Africa rain brings a flush, not a dearth, of insects to the land.

> As in a drought the thirsty creatures cry,
> And gape upon the gathered clouds for rain;
> And first the martlet meets it in the sky
> And with wet wings joys all the feathered train.
>
> John Dryden in *Annus Mirabilis* (1667)

Only in Africa, which he did not know, could Dryden's pleasant conceit have been true. In Damaraland, for instance, common swifts appear irregularly during the winter, chiefly after rain. In the Belgian Congo they are seen mainly on rainy days, when they descend low to feed on the swarms of flying termites or flying ants. It was after rain in Kenya that, as mentioned in Chapter 1, nine different forms of swifts were found in one flock. In a recent letter from Northern Rhodesia, C. W. Benson writes that, now that the rains have come, common swifts have appeared in great numbers to feed on the flying ants and termites. Many similar quotations could be given, but these are enough to show that the swift's appearance with rain is for a very different reason from that which holds in northern Europe.

Swallows show various resemblances to swifts, and in Britain swallows and sand martins are sometimes seen flying in large numbers against the wind in unsettled weather, but their movements have not been analysed, so it is not known whether they are comparable with those of swifts. With this possible exception, no other type of bird is known to make spectacular weather movements like the swift, but then few other kinds of birds are deprived of their food by rain, and extremely few others would have the power to travel rapidly for long distances to avoid bad weather.

Chapter 14

MIGRATION

One of the few literary allusions to the swift concerns its migration, though through a mistranslation it has been hidden from English readers. In the eighth chapter of Jeremiah, the Authorised Version reads that 'the stork in the heaven knoweth her appointed times; and the turtle and the crane and the swallow observe the time of their coming'. Here the stork (*hasidah*) and the turtle dove (*tor*) were translated correctly, but the other two birds, *sŭs* and *'agur* in the Hebrew, were not. *Sŭs* is now thought to mean a swift, the word being used in this sense in modern arabic. Moreover the swift is a prominent day-flying migrant through Palestine at its appointed time and vanishes for the winter. The swallow, on the other hand, fits the context less well, as many stay for the winter in the Holy Land, while there is another Hebrew word, *deror*, for the swallow, as used in Psalm 84. Alternatively of course, *sŭs* might have been used loosely for both swifts and swallows. It occurs in one other place in the Bible, in the 38th chapter of Isaiah, where it is again linked with the *'agur*, both being said to chatter. Such a verb might perhaps be used for the scream of the swift. The bird meant by *'agur* is unknown, and the various published suggestions do not fit for a noisy day-flying migrant. Another species which migrates conspicuously by day through Palestine and has a noisy call, but is not otherwise mentioned in the Bible, is the bee-eater. Could this have been the *'agur*?

The first swifts usually reach England near the end of April, the main body in the first three weeks of May. Precise dates for the return of the resident birds were obtained by visits to the tower each night in May during four years. In 1950, which was perhaps an average season, the first swift came on 1 May, and most arrived between the 4th and 17th of the month. In 1955 they were slightly earlier, nearly all arriving between 1 and 14 May. In 1949, on the other hand, they were decidedly later, the first appearing on 5 May and most between 11 and 27 May. In these three years the colony assembled gradually, with two to five newcomers each day and an occasional bigger arrival. The largest arrivals, involving nearly a quarter of the whole colony, were on 22 May 1949 and 6 May 1955.

In 1951, the pattern of arrival was different. The start was normal, the first two birds coming on 1 May and thirteen more during the next five days, up to 6 May. Then there was a hold-up, and while those that had already arrived stayed with us, there were no newcomers for eight days. The next did not come until 15 May and the rest came in gradually over the next nine days. In this year, members of the British Trust for Ornithology were watching for migrant swifts and found a hold-up between the same dates. The weather maps show that for a few days up to and including 6 May, the wind over France was southerly. On the following day, when the hold-up began, a depression moved in eastwards from the Bay of Biscay, while on 8 May a north-easterly air-stream spread over France. From then until the last day of the hold-up on 14 May, the weather in France was cold with strong north-easterly winds. On 15 May, when fresh birds arrived at Oxford, the wind at last changed to the west over most of France and northern Italy, and though it was still northerly over the extreme north of France and over England, the swifts would have had to make only the last stage of their journey against a cold wind.

The autumn departures have also been studied at the tower in four years. The young swifts, as already mentioned, fly from the nest independently of their parents, when their wings have reached full, or nearly full, length for the time being. In each year they have left fairly evenly over about a fortnight, with no tendency to depart in waves or groups. Most have gone in the last few days of July and the first few days of August, our earliest date being 19 July and our latest (from a repeat clutch) 7 September.

Contrary to what has sometimes been stated, the parent birds usually remain for several days after their brood has gone, but the interval has varied greatly with the weather. In the extremely wet July and August of 1954, the parents stayed for an average of 11 days after their brood had gone, one pair staying as long as 26 days. In rather poor weather in 1952 the average interval was 8 days, in mainly fine weather in 1951 it was 3½ days, and in extremely fine weather in 1955 it was only 2 days. In 1955) almost all the parents left within a day or two of their young except for two pairs which remained for another 15–16 and 10–11 days respectively, thus considerably raising the average. The pair which stayed longest was in the same box as the pair which stayed longest in the previous year, and presumably the same individuals were concerned.

In Switzerland, Weitnauer found that the parents normally migrated on the same day as their young, thus carrying further the tendency found in fine summers at Oxford. Indeed, one member of the pair sometimes left several days before the brood had gone, and this was also found at Oxford, in over a quarter of the pairs in the two fine summers, but rarely in 1952 and not at all in the wet summer of 1954. Usually, one partner left in this way only if the brood consisted of a single nestling, or after all except one of a larger brood had flown.

It may be suggested that the parent swifts tend to stay for some days after their brood has flown in order that they can put on fat and get into good condition for their long journey. It fits with this view that they stayed back for a much shorter time in the fine than the wet summers, as when food is plentiful, they can feed their young more efficiently and themselves put on weight more quickly. Further, any pairs which lost their eggs and so had no young to feed might be expected to be in better condition than those with young, and in fact such pairs departed on the average several days earlier than those with young, especially in fine summers. Migratory birds normally put on fat before setting off, to act as a reserve on their journey, and swifts are no exception. In midsummer the adult weighs about 43 grams, but a migrant in good condition weighs 54 grams, about one-quarter as much again. Weight for weight, fat is a more efficient fuel than high-octane petrol. That it may be needed is shown by the fact that three adult swifts that came to rest on the lighthouse at Öland in Sweden, at the end of a heavy day's passage against the wind, weighed only 35 grams by the following morning. If they had set out weighing 54 grams, they must have lost over one-third of their weight.

The non-breeding pairs, mostly yearlings, left with the main body of adults but chiefly near the end, so that their average date of departure was several days later (except in the very wet year, when they left at the same time). It might have been supposed that they, like the adults which failed to hatch their eggs, would have got into good condition ahead of parents with young, but either this was not so or they defend their nesting sites for as long as possible. After the main body of adults has gone, only a few parents with late broods remain. Charles Darwin and several later authors claimed that such birds desert their young when the rest of the colony has left, but this has not been our experience. They usually left on the same day as their young, though one of them (never both) occasionally left a day or two beforehand. In the unusually warm autumn of 1835, a pair of swifts is reported to have stayed to raise young which flew as late as 4 October.

Although in each year the departures of the adults were spread over two or three weeks, the two members of one pair usually left within a day of each other. When one adult left before the brood had gone, the other waited for its young, but excluding these pairs, the two adults left on the same day in a quarter of the pairs and a day apart in half the pairs. Most of the rest left within four days of each other, but two after a week and one after a fortnight. After its mate has gone, the remaining bird seems very restive in its box at dusk, particularly on its first evening alone, and this may be why it so often leaves on the day after its mate.

In 1951 and 1952, most of the adults left during one of three waves of departure, each of which lasted two or three days, and few left on intervening days. In the fine summer of 1955, on the other hand, departures were spread out very evenly, from two to nine individuals leaving each day between 30 July and 16 August. There were only two days when none departed, 9 and 10 August, on both of which the sky was completely overcast. In the wet summer of 1954 the picture was different again, a small number leaving between 2 and 13 August, and three-fifths of the whole colony during the next three days, twenty-one of them on a single day. As a result, the average date of departure was six days later in 1954 than in 1955, though the last bird left only a day later. In 1954 the weather remained bad until the end, and it seems unlikely that the adults could have found enough food to put on extra fat. That, nevertheless, all had gone by about the same date as in other years suggests that there is a limit to the time for which they are prepared to stay back. This is supported by the fact, already mentioned, that parents with late broods leave immediately after their young and do not remain for even one more day.

In 1951 the second wave of departures ended on 9 August. There was then a gap of seven days before any others left. This was probably due to the weather, as on the first five days it was extremely bad and the birds could have collected little food. There followed two lovely days, on which we thought the birds might have left, but they did not do so, presumably because they were feeding hard. The early morning of the next day, 17 August, was overcast and rain started at 7.45 a.m., lasting the rest of the morning. From 7 a.m. onwards we heard the excited screaming of swifts flying round the houses of

Oxford and by 8 a.m. a flock of a hundred had collected, presumably derived from a number of different colonies. The birds circled up and gained height and then most of them set off about SSW, rising as they went, and vanished into the low clouds. Some dropped out and returned, joining up with another party of about fifty birds, most of which left in the same way a few minutes later. That evening, the nightly check at the tower showed that seven out of the nine remaining adults had gone.

I watched another departure on 6 August 1955 from outside the tower. The weather was fine with a mild north wind, but there was thick mist until 7 a.m. and the sky was wholly overcast for more than an hour after this. At 7.45 a.m. about thirty swifts were flying round the tower, often following each other in a typical slow-flying screaming party. At 8 a.m. most of the screaming stopped, the birds circled higher, to some fifty feet above the tower, and five minutes later ten of them went off silently SSE, circling and gliding at about two hundred feet above the ground. All, or nearly all, of these birds returned a few minutes later and joined the others round the tower, and screaming was resumed. Shortly afterwards the birds again rose higher, another small party detached itself, set off and again returned, while at 8.18 a.m. nine went off SSE and were not seen to come back. From then on, only fifteen birds were round the tower and there was less screaming. After a visit to the nests, I came out again at 8.40 a.m., and shortly afterwards ten swifts flew silently high above the tower. I formed the impression that they were about to migrate, but was much puzzled when they set off in a definite manner northwards. They rose higher, circling, and as I followed them through field-glasses I saw high above them in a gap in the clouds a circling party of about 150 swifts. This large group was invisible to the naked eye. The nine from the tower joined them and all drifted south, soon disappearing into the mist. At 9 a.m. another fifteen birds left the tower SSE in the same manner as earlier on, and ten minutes later twenty-five did so, but ten of these returned. Soon these also went off. That night, five individuals proved to have left the colony, much fewer than I had expected. It suggests that those round the tower were joined by swifts from other colonies, and also that many of those leaving SSE, as if migrating, returned again later.

Regular watch was kept during the next seven mornings, and except that I saw no more large parties high overhead, the procedure was the same as that just described. First there were screaming parties, then higher flights above the tower in silence, in which birds from other colonies probably joined, then departures of between five and twenty birds, drifting off rather east of south, and often returning. On the two overcast mornings, when the nightly check showed that no birds left, behaviour was similar to that on other days except that there was less screaming and perhaps more false starts southwards. The observations suggest that, at this time of year, feeding swifts tend to leave the tower southwards, like migrants, during the morning. As already mentioned, there were no big waves of departure in 1955, so that the excited exodus seen four years earlier was not repeated.

Shortly after the departures had taken place on 17 August 1951 and 6 August 1955, I watched for migrating swifts on the hills near Oxford, but saw none, though on the second occasion the sky was clear. Watch was kept on two other days in 1955 on which swifts were known to have left, but in several hours I saw only two parties, of five and twenty individuals, drifting south over the Berkshire Downs. On these three days in August 1955, the sky was clear and the wind northerly, and it is reasonable to presume that most swifts were migrating at too great a height to be visible from the ground. In support of this suggestion, on 26 July 1917, when Professor Stresemann was in an observation balloon on the Western Front, he saw eleven swifts flying south at about midday at a height of over 2,000 feet from the ground. Likewise a French pilot, at the time of the autumn migration, saw two pairs of swifts flying in bright sun above dense cloud at 7,000 feet in the Lyonnais. Again, in the Pyrenees near Andorra on 20 August 1950, a fine day with a north-easterly wind, an English observer watching for migration saw no swifts until 4 p.m. Then, clouds formed on the mountains and came down on the Pas de la Casa at 6,800 feet, and an immense flock of swifts appeared and circled below the cloud near the head of the pass. Presumably, swifts had been migrating during the day, but too high to be seen until they were brought lower when cloud hid the way ahead.

There are other records, at various times in the summer, which show that swifts sometimes fly too high to be seen from the ground. British pilots have reported single swifts at 2,500 feet, 4,300 feet and 7,500 feet and a flock of twenty at 3,400 feet, and individuals or parties have been seen at about 6,000 feet over Holland, Switzerland and Iraq. The wind is usually stronger high up than near the ground, so that with a following wind it is reasonable that swifts should migrate at a height. At other times they fly lower. Thus on a drive from Oxford to Northampton in early August, I passed several flocks of swifts, each consisting of perhaps fifty to a hundred individuals, drifting southwards in spirals in the usual way, but unfortunately I did not record the weather at the time. Swifts do not migrate low over the ground except in heavy rain or with a strong headwind, when they beat steadily forward instead of circling and gliding. Probably they do not start on migration under such conditions, but they may meet them on their journey. A spectacular passage of this nature was reported in May 1949 from the Camargue and the Massif Central in southern France, when thousands of swifts were seen struggling northwards close to the ground against an extremely strong headwind, under conditions in which most other migrants did not attempt to travel, and honey buzzards were blown backwards in the gusts.

The evidence suggests that in the autumn swifts have set off from the tower between 7 a.m. and 9 a.m., but they may also leave at other times of day. On two fine and clear mornings in 1954, after a long spell of bad weather, the birds almost certainly left before 7 a.m., while on one day in 1955 there was an excited party round the tower soon after noon, the behaviour of which strongly suggested that they were about to set out. Moreover, on two days in 1955, individuals present in their boxes at noon had gone by nightfall. A Dutch worker has stated that migration regularly starts in the evening, but I should like further evidence of this. It is not known whether swifts migrate at night. A number have been found dead below lighthouses, but these were probably birds roosting there after a heavy day passage and not birds attracted to the light during a nocturnal movement. As swifts may spend the night on the wing, there is nothing improbable in their migrating at night, but direct evidence for this is lacking.

Swifts arrive later than most of our summer visitors, though the spotted flycatcher and red-backed shrike usually come a few days later and the marsh warbler and quail about a month later. The date in spring on which the first swift has been seen at Oxford, averaged over thirty-five years, has been 29 April, as compared with a day earlier in Cheshire and a day later in Hertfordshire. On the east coast, the birds seem to come rather later, the average date in Norfolk and Suffolk being between 4 and 6 May. On the Continent, the average date of arrival varies with latitude, as would be expected, the average being 18 April in Madrid and 20 May in south-western Finland.

At Oxford, as already mentioned, most of the adult swifts have left by the third week of August. In both Spain and Switzerland they leave earlier, before the end of July, a point which puzzled Gilbert White and several later writers, since in late summer the weather is much more favourable for swifts in these warmer countries than in England. The reason, I suggest, is merely that the birds depart as soon as they have raised their young and are in fit condition to migrate, and that they are not ready so early in England as in Spain or Switzerland. For the same reason, those breeding in Scandinavia leave after the English birds have gone. Most of the swifts seen in England in late August and September are probably birds on their way south from Scandinavia.

> By nature taught to know their stated time,
> They leave, without delay, the favorite shore,
> And wait content, in some more genial clime,
> Till spring, returning, calls them back once more.
>
> Thus still may Nature's voice new lessons give,
> For she in all her works some hint supplies;–
> Either to teach us here below to live,
> Or to prepare us for our native skies.
>
> Rev. R. Hennah, in *Truth's Mirror or friendly hints to young persons* (early nineteenth century)

Swifts spend the winter in South Africa, but the route by which they reach it is not certainly known. In the hold-up of migration in May 1951, mentioned earlier, swifts arrived at Oxford on the day that the wind changed from north to west in central and southern France and northern Italy. At this time the wind was still northerly in Spain, which suggests that the birds came through Italy and eastern France and not through Spain and western France. In support of this idea, a swift that we ringed near Oxford as a breeding adult in 1947, and which we caught again at its nest in the three following years, was reported dead on 3 June 1953, at Pontarlier, Doubs, in the extreme east of France. This bird was presumably on its spring migration, the late date being due, perhaps, to its having died some time before it was reported. It might, however, have been deflected from its normal course by bad weather. There is also a curious record of a swift found dead in late May 1886, at New Ross in Ireland, bearing a paper on its leg with the words: 'Mary Elsam, Suakim, Egypt, 10-3-86.' The date and place (presumably Suakin on the Red Sea) seem reasonable and would favour the idea of an easterly route in spring, but no one seems to have thought it worth while writing to Mary Elsam, so a practical joker cannot, perhaps, be excluded.

The recoveries of swifts ringed abroad throw little light on their migration. Few have been found outside the country where they were ringed and most of these were in an adjoining land less than two hundred miles away, where the swifts might well have travelled on a weather movement. Of the recoveries at a greater distance, a bird ringed in Sweden one summer was found in the following spring in Jugoslavia (Servia), indicating a south-easterly migration route, which many other species of Swedish birds are known to take. Autumn recoveries include one ringed in Sweden which was found to the south in Czechoslovakia, two from Germany and one from Czechoslovakia found in Spain, indicating a south-westerly route, and one from Holland found in Austria, indicating a south-easterly route. These are too few on which to base any conclusions. The most spectacular recoveries have been of two ringed in Germany and one ringed in Switzerland which were later found in the Belgian Congo, and another ringed in Germany and found in French Equatorial Africa.

The status of the swift in Africa is incompletely known, partly through the difficulty, if not impossibility, of distinguishing it in the field from the resident African black swift (*A. barbatus*). Reliable records therefore depend on collected specimens, but swifts are hard to shoot and many collectors leave them alone, so that there are comparatively few in museums. Common swifts have been taken in the Union of South Africa in the four months of November to February inclusive, and in the same period, with the addition of a few in October and March, in Northern Rhodesia and Nyasaland.

All the specimens collected in December and January have come from Africa south of latitude 10°S, which is probably the main wintering ground of the species, though as yet less than twenty specimens have been taken there. Further north, in the Belgian Congo, common swifts occur chiefly between August and October, with a few in November, and again between late February and April. It seems likely that the same holds in East Africa, but there are too few records to be sure. If the term winter is taken to include November and February, common swifts range from equatorial to South Africa; it is only in midwinter that they seem confined to Africa south of 10°S.

Common swifts from Europe and Asia mingle in Africa, and so far as is known none spend the winter in tropical Asia. Those breeding on the Pacific coast of China have a long journey, west across the length of Asia and then south to South Africa, but there are several other species with weaker flight that breed in Asia and make a similar journey, such as the willow warbler. Wheatears move even further, since those breeding as far east as Alaska spend the winter in Africa, travelling about 7,000 miles twice each year.

Most common swifts leave Africa south of the Sahara for the summer, but a number have been collected in June out of big mixed flocks of swifts in Darfur in the southern Sudan. These birds were far to the south of the known breeding range, they had under-developed sex organs, and were presumably non-breeding yearlings which had not completed their northward migration. Pallid swifts were collected with them in the same condition.

Nearly all those song-birds which breed in Britain and migrate to Africa for the winter renew their feathers by a full moult in the late summer before they leave, but swifts and swallows leave Britain almost immediately after they have raised their young and undergo their main moult in Africa. Curiously, the alpine swift stays on in Switzerland after it has raised its young and moults before leaving in late September and early October, two months after the common swifts have gone.

Alpine swifts from Europe spend the winter in equatorial Africa,

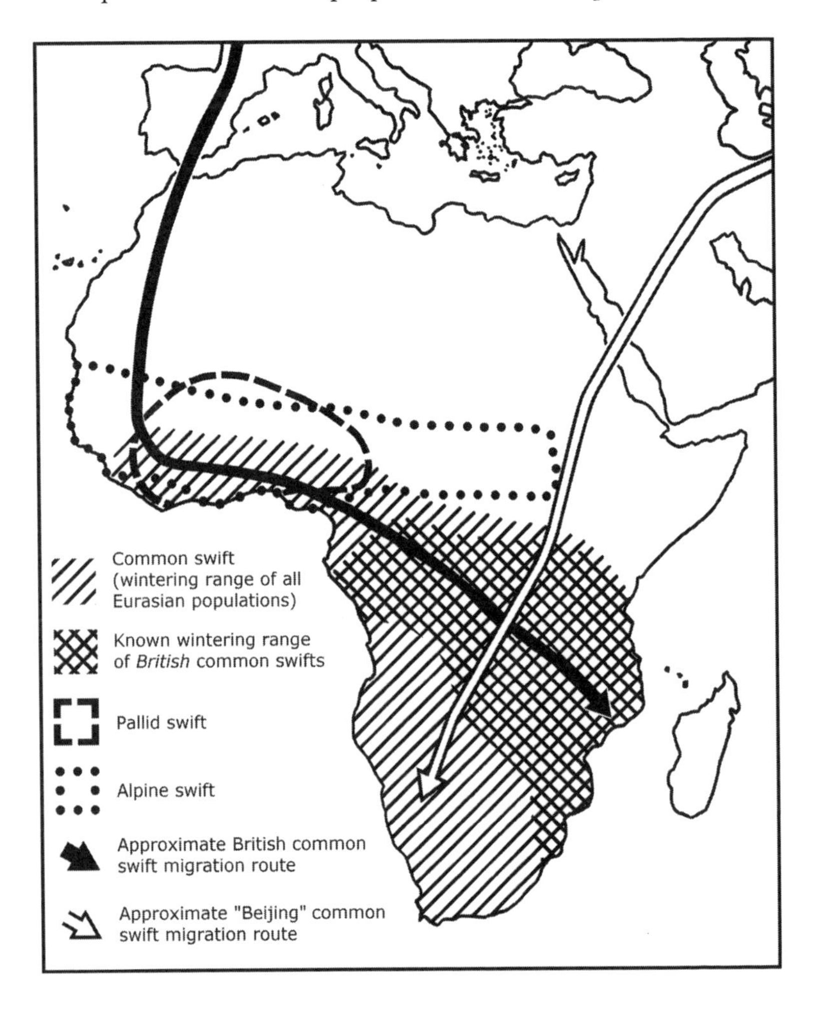

Fig 18: Winter range in Africa of common swift (*A. apus*) (midwinter only), alpine swift (*A. melba*) and pallid swift (*A. pallidus*)

having been collected at this season in the extreme east of the Belgian Congo and adjacently in the southern Sudan. They do not, so far as is known, move further south, and reports of them from SW Africa are wrong, due to confusion with a local resident form. An alpine swift ringed in Switzerland has been found 900 miles to the southwest in Spain on its spring migration and six others have been found from 170 to 380 miles to the southwest in France in spring and autumn, but none have been reported to the southeast. This suggests that the Swiss birds travel between Europe and Africa by a south-westerly route.

The other European species, the pallid swift, has been found in winter in Darfur in the southern Sudan and also at Aden in southern Arabia (in both of which places some also stay the summer without breeding). In addition, one specimen has been collected in Somaliland and another in Uganda, but it is hard to know how much importance to attach to these isolated records. Reports of pallid swifts from further south in Africa are wrong, due to confusion with other species.

Summarising, of the three species of swifts breeding in Europe, the one with the most southerly breeding range has the most northerly winter quarters, while the one which breeds furthest north in Europe moves furthest south in Africa. Thus the breeding places of the pallid swift on the northern shores of the Mediterranean are some 2,000 miles from Darfur and Aden, but alpine swifts have to travel about 3,500 miles from Switzerland to the Congo, and our common swifts at Oxford move over 6,000 miles to reach South Africa. Existing knowledge on these points, which may be very incomplete, is summarised in Fig. 18.

The other swifts breeding in the North Temperate region also migrate south for the winter. The Pacific white-rumped and needle-tailed swifts of northern Asia spend the winter in the Australasian region, while the Vaux and black swifts of western North America travel to Central America. The winter home of the chimney swift of eastern North America was for long unknown, despite the huge number caught and banded at large chimneys where they roost on their way south in autumn; but several bands have now been returned by primitive Indians living in the remote forests in the interior of Peru. The distribution of the many swifts breeding in the tropics is too imperfectly known to state whether any of them migrate, but in Venezuela, as mentioned earlier, several species have to come to lighted windows on misty nights, which suggests that they may have been on migration.

Chapter 15

THE RACES OF SWIFTS

This chapter represents an interlude in the main story, and the work for it was done in London instead of Oxford. As mentioned in the last chapter, some of the alleged specimens of European swifts collected in Africa were wrongly identified, really belonging to local African forms. Swifts, like most other birds, are divided into geographical races or subspecies, which differ from each other in various ways. Confusion arises if two races of the same species from distant areas look very alike, or if a race of one species looks very like the members of another species; and both sources of confusion occur in swifts.

A simple instance concerns the alpine swift (*Apus melba*). This grand bird, the largest in the genus, is brown on the upper parts and differs from all the other species in being mainly white below. As its English name suggests, it provides one of the ornithological attractions of Switzerland, in which connection it is told that in the Edwardian heyday the great Lord Rothschild was once sitting with his colleague Dr Hartert on a hotel terrace at St Moritz when he half rose in excitement exclaiming: 'Ach, there's Melba!' The fashionable crowd turned, hoping to glimpse the great singer. 'Are you sure?' said the cautious Hartert. 'Yes,' said Lord Rothschild, 'I recognised her by her white belly.' The white underparts, interrupted by a brown band across the chest, are diagnostic, and mean that, however great the racial variations, forms of *A. melba* can reliably be distinguished from all the other species of *Apus*.

The alpine swift breeds not only in southern Europe but also in southern Asia from Palestine to India and in much of Africa. In South Africa it is called the great swift. Over this wide range it

Fig 19: Alpine swift (*Apus melba*)

varies somewhat in appearance, especially in the colour of the upper parts and in size. Most of the races are brown on the upper parts, like the European form, but the shade of brown varies somewhat from place to place, being paler in drier and darker in wetter areas. At one extreme, there are two widely separated forms, in the desert regions of Somaliland in eastern Africa and Damaraland in Southwest Africa, which are a pale greyish-brown. At the other extreme, two forms from areas of high rainfall, Ruwenzori and Madagascar respectively, are as black above as common swifts. In size, the birds from Ruwenzori are the largest (with a wing-length of 23 cm) and those from Madagascar are the smallest (with a wing-length of 19 cm), and neither of these forms overlaps appreciably in measurements with the birds from other parts of Africa. When both colour and size are taken into account, it is possible to identify with confidence specimens of the alpine swift from various parts of the world, eight distinct races being separable in this way, while within some of the widespread races occur smaller but definite variations which have not been thought sufficient to warrant separate naming. Bird watchers have sometimes felt that such minor differences are not worth serious attention, but apart from their adaptive value to the birds, considered later in this chapter, they help in the study of migration. Thus although the alpine or great swifts breeding in Southwest Africa look very like those from southern Europe in colour, they can reliably be separated by their smaller wing. As a result, the alleged specimens of the European form from this area can be proved to have been wrongly identified, as all have the shorter wing of the local African form.

One other species, the pallid swift (*Apus pallidus*), breeds in Europe, occurring in Spain and Dalmatia, while by chance we found the first pairs nesting in France, on houses in the seaside resort of Banyuls – as if one should find a new species for Britain breeding on the front at Brighton. The pallid swift is extremely like the common swift, but is brown not sooty on body and wings and has a pale forehead. Museum specimens can also be recognised by the narrow pale tips to the feathers of the underparts, the shallow fork to the tail, and by the first two primaries being about equal in length (whereas in the common swift the second primary is about 5 mm longer than the first). These are small differences, but though the common and pallid swifts look so similar they are distinct species, they do not intergrade in their characters and there is no suggestion that they ever interbreed, though there are places where they breed side by side.

If the uninstructed reader were shown specimens of the alpine swift from Ruwenzori, where they are large and dark, and from Somaliland, where they are small and pale, he might well suppose them to be different species. On the other hand, specimens of the common and pallid swifts look so alike that he might well suppose them to be races of the same species. This type of difficulty occurs in various other swifts, and as a result it is often hard to be sure of the correct relationships of different forms. In general, however, distinct species differ from each other in a number of features, even if to only a small extent in any one of them, and further they do not intergrade in their characters. Races of the same species, on the other hand, even when they look very unlike, usually differ in only one or two ways, though these may have a marked effect on the appearance of the bird. Thus a general darkening of the plumage can cause a marked difference, though basically due to a single character. In addition, races often intergrade with each other in appearance and their variations usually follow regular trends. Colour, for instance, tends to be paler in drier and darker in wetter areas, while size tends to be smaller in hotter and larger in colder regions.

These points help to clear up an apparently simple question, namely the extent of the range of the common swift. That it breeds through Europe and much of Asia is generally agreed, the Asiatic race (*A. apus pekinensis*) being browner than the European race (*A. apus apus*) and having a more extensive pale throat, but obviously belonging to the same species. The doubt arises in Africa, where forms of the common swift have been claimed to breed in both the east and the south. The forms in question differ in various ways from our bird, and the problem is to know whether they are races of it or whether they really belong to different species. To pronounce on this question, it was necessary to review all the species and subspecies of swifts in the genus, a type of research rarely undertaken by the modern bird-watcher, though it was a favourite pastime with the

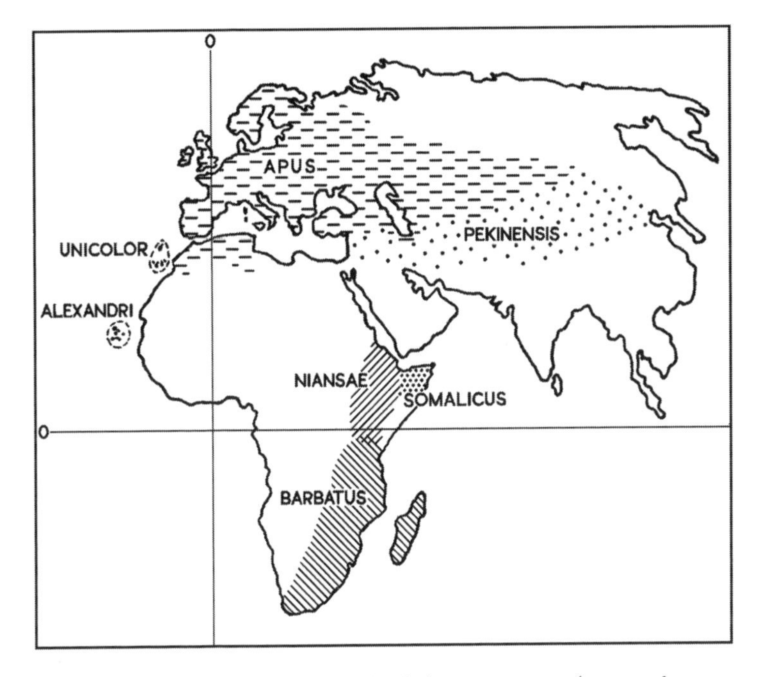

Fig 20: Races that have been claimed to belong to *A. apus* (*apus, pekinensis, unicolor* and *alexandri* do so; *niansae, somalicus* and *barbatus* do not)

older naturalists at a time when many of the world's birds had still to be recognised and named. It requires handling museum skins in London, which is not nearly so pleasant as watching birds in the wild, but there is satisfaction in straightening out a tangle of names, particularly where, as in the present instance, there was much confusion and wrong identification. More important, after this sorting out, underlying adaptations were revealed which were previously unsuspected.

The common swift certainly breeds as far south as Northwest Africa. There is then a gap in the range due to the Sahara desert. Further south, the latest official check-list of birds of the world shows two more races, Shelley's swift (*niansae*, formerly called *shelleyi*), which breeds in Abyssinia, Eritrea and Kenya, and the African black swift (*barbatus*), which breeds from western Kenya south to the Union of South Africa. The ranges of these forms are shown in Fig. 20. Actually both Shelley's swift and the African black swift breed, without interbreeding, at Nakuru in Kenya, so they must belong to different species, and at most only one of them could be a race of the common swift.

Shelley's swift is a dark sooty brown, like the Asiatic form of the common swift, but it is much smaller, has narrow pale tips to the feathers of the underparts, a shallower tail-fork and the first and second primaries about equal in length. It is extremely like another small swift (*somalicus*) breeding adjacently in Somaliland, except that *somalicus* is brown instead of sooty in colour, and in my view these are certainly races of one species. This Somaliland form is normally considered to be a race of the pallid swift, which it closely resembles except in being smaller, and this view I accept, which means that Shelley's swift must also be treated as a race of the pallid swift. In the minor points in which Shelley's swift differs from the common swift it resembles the pallid swift, and although it is darker, this fits with its living in a wetter area.

The African black swift is harder to place. It looks so like the common swift in colour and size that the two are not normally distinguished in the field where they meet in South Africa in winter. If they belong to one species, its range is highly unusual, with a gap of two thousand miles where neither breeds, between Northwest Africa and the Equator. Such a gap fits better with the idea that they are separate species. Further, a close examination of museum skins shows that the African black swift differs from the common swift in quite a number of small ways. The feathers of the mantle (upper back) are glossy black and are in marked contrast to the bronze-coloured quills, whereas in our bird both mantle and primaries are sooty.

The dark feathers of the underparts have narrow pale tips, which are lacking in our bird. The pale throat is more extensive, and these feathers often have dark shafts, which are not found in our bird. The first and second primaries are about equal in length, instead of the second primary being longer, and the tail is decidedly less forked. Any one of these differences taken by itself seems trivial, but a number of separate features are involved, and the differences are of the same order as those separating the common from the pallid swift, which are known to be different species since they breed in the same area without interbreeding. I therefore consider the African black swift to be a separate species from our own, although it looks very like it.

While I have thus removed two African forms from our own species, I have incorporated with it two others which in the latest check-list were placed in another species. In Madeira and the Canary Islands occurs another black swift (*unicolor*), very similar to our own except that it is decidedly smaller and has a darker throat. Nearly all the birds breeding in Madeira and the Canary Islands belong to European species, so it would be strange if this did not also hold for the Madeiran black swift. Further, the dark throat does not provide an absolute distinction from the common swift, paler individuals from Madeira matching darker individuals of the common swift from Britain. I therefore consider the Madeiran black swift to be a small race of the common swift.

Further south, in the Cape Verde Islands, occurs a small brown swift (*alexandri*), which is very similar to the Madeiran bird except that it is yet smaller, and brown instead of black. All recent authorities have regarded it as a race of the Madeiran black swift, and with this I agree, which means that it should now be treated as another race of the common swift. It may be added that although the Cape Verde Islands are fairly far south down the African coast, a number of European species breed there, including the familiar blackcap.

As already mentioned, the only conspicuous difference between the common and pallid swifts is that the one is sooty and the other brown. If the foregoing views are accepted, then in one race of each species the colour difference is reversed, since the Cape Verde race of the common swift is brown and the East African race of the pallid swift (i.e. Shelley's swift) is sooty. This is at first sight surprising, but the variation is in line with well-established trends, since Shelley's swift is darker than other forms of the pallid swift and breeds in a wetter area, while the Cape Verde birds are paler than other races of the common swift and breed in a drier area. Fortunately, colour does not provide the only difference between the common and pallid swifts, and the presence or absence of pale tips to the feathers of the underparts, the extent of the tail-fork and the relative lengths of the first and second primaries, provide supporting evidence for the views adopted here. Further, each of the exceptional forms breed near to another race which bridges the gap with the typical form of the species, the Madeiran black swift being intermediate in appearance between the Cape Verde swift and the European common swift, and the Somaliland form between Shelley's swift and the Mediterranean forms of the pallid swift. Clearly, however, a problem of judgement is involved, especially as the available evidence consists of museum skins and not of living birds.

The critical point is to determine which differences between two forms are significant and which unimportant in relation to their affinities. Two forms may look very alike either because they are closely related or because they have independently evolved similar features as an adaptation to the same way of life. This, of course, is the basic difficulty in any classification in which the animals are placed in an order denoting their natural relationship. When two species are remotely related, it is easy to recognise resemblances between them that are secondary, due to convergent adaptation. Thus only the most naive person would now classify the bats with birds rather than mammals, though they were put with birds in sixteenth-century works. When the species are less distantly related, the problem is harder. It needs a fuller knowledge to be sure that hawk and owl, or horse and cow, are not very closely related, and that some of their more obvious points in common are convergent adaptations to a similar way of life.

When the forms in question are closely related, it becomes exceedingly hard to tell which of their resemblances are due to common descent and which to convergent adaptation. Thus, if a Martian explorer returned to his planet with badly stuffed skins of a gorilla, a Zulu and a Viking, the local museum workers might well classify the gorilla and the Zulu in one species and the Viking in a different one. In this, they would have been misled by a superficial character (colour), combined with the many features shared by higher ape and man through their common descent. Internal anatomy and behaviour would, of course, show at once that Zulu and Viking are conspecific and the gorilla highly distinct. Now the species in the genus *Apus* are much closer to each other than is man to a higher ape, added to which their behaviour and internal anatomy are largely unknown. It is not, therefore, surprising that they are hard to classify.

The most striking instance of racial variation occurs in the Pacific white-rumped swift (*Apus pacificus*) which breeds in Siberia, China and Japan and is distinguishable from other swifts by its white rump, well-forked tail and barred underparts. To the south of it, in the Khasia Hills in northeast India, occurs a smaller and darker bird (*acuticaudus*), with a similar well-forked tail but no white on the rump. It also differs in having glossy bluish-black instead of dull sooty upper parts, narrower pale bars on the underparts and dark streaks on the white throat; and hitherto it has been considered a separate species.

In Burma occurs an undoubted race (*cooki*) of the Pacific white-rumped swift. As shown in Fig. 21, its range adjoins that of the typical form, with which it agrees in its essential features, including the white rump. But it resembles the Khasia Hills swift in its small size, in being glossy blue-black on the upper parts, with narrow pale bars on the underparts and dark streaks on the throat. Further, its white rump is only half as large as that of the typical race, and shows dark streaks. Hence this Burmese form bridges the gap in appearance between the Pacific white-rumped and Khasia Hills swifts, and its

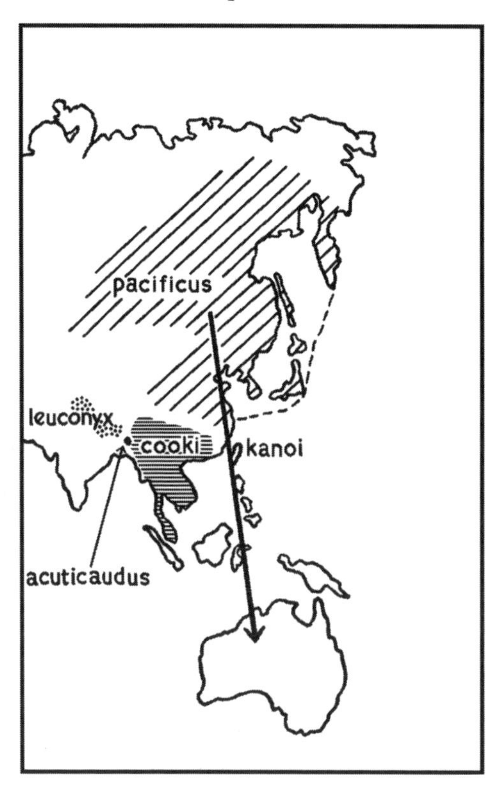

Fig 21: Distribution of Pacific white-rumped swift (*A. pacificus*) showing the range of the five races (two of which are mentioned in the text), and the migration of the northern race to Australia

range adjoins theirs. Moreover, the darker colour of the Burmese birds fits with their living in a wetter area than the typical form. Now the Khasia Hills are credited with having the highest rainfall of anywhere in the world. In view of this evidence, it is reasonable to suppose that the Khasia Hills swift is not really a separate species but an extremely dark race of the Pacific white-rumped swift. Although it looks so different from the typical form, especially in the absence of a white rump, almost all the differences depend on a single variation, the darkening of the plumage.

A similar type of racial variation, but in reverse, is shown by Bradfield's swift (*bradfieldi*) from Damaraland in Southwest Africa. This is a brown bird of about the size of a common swift. The first collected specimen was wrongly supposed to be a pallid swift in winter quarters, but it can reliably be distinguished from the forms of the pallid swift by its longer wing. Later it was described as a new species, but since it is confined to Damaraland, it is more likely to be a local race of some widespread species. It was then put with the African mottled swift (*Apus aequatorialis*), one of the few brown-coloured swifts in southern Africa, but this was a mistake, as it differs markedly from it in several respects. Yet later it was thought to be a large race of the pallid swift, though the latter does not otherwise breed within two thousand miles. Now, except in being brown instead of black, Bradfield's swift is extremely similar to the African black swift, and I consider it to be a pale race of it from an unusually arid area.

The tendency for races of the same species to be darker in wetter and paler in drier areas is widespread in birds and other animals, and forms part of Gloger's rule (the other part being that animals tend to be paler in colder and darker in warmer parts of their range). In some species of swifts the variations due to Gloger's rule are small, and even large differences may cause no confusion when, as in the alpine swift, the species possesses one or more characters which readily distinguish it from all others. But when two species, such as the common and pallid swifts, look very alike, then marked variations due to Gloger's rule may lead to greater differences in superficial appearance between subspecies of the same species than between full species. Thus examples have been given of a brown race of a species that elsewhere is black, of a sooty race of a species elsewhere brown, of a glossy blue-black race of a species elsewhere sooty, and of a dark-rumped race of a species which elsewhere has a white rump. To recognise such cases, detailed examination is needed of the small and apparently trivial differences between species, of their geographical ranges, and of the rainfall to which each form is subject.

Although it may seem surprising that two races of the same species should look less like each other than two full species, the European bird-watcher can see a parallel example at home. In England the willow and marsh tits look so alike that most naturalists prefer to identify them by their calls rather than their appearance. But the Scandinavian race of the willow tit is paler, greyer and larger than the British race, with the result that the British willow tit looks more like another species, the marsh tit, than like the northern race of its own species.

That Gloger's rule holds true in so many kinds of birds and other animals suggests that it has general adaptive significance. Probably, this is connected with the loss of heat from the body. As is well known to the householder, radiators give out more heat when painted black than white. Likewise, to maintain the same degree of heat, an animal will need to be paler in the colder than the warmer parts of its range. Also, since heat is lost by evaporation more quickly in a dry than a wet atmosphere, the animal will need to be paler (i.e. to give out heat less easily) in the drier than the wetter parts of its range. (In certain animals, racial variations in colour probably help in concealment, but as swifts are usually either in flight or at rest in holes, they do not need concealing coloration.)

In swifts, as in many other birds and mammals, each species tends to be larger in the cooler than the warmer parts of its range (Bergmann's rule). This also is thought to be an adaptation for conserving heat, since a larger animal has a proportionately smaller surface from which to cool off. In five species of swifts, the tropical races are so much smaller than those from temperate regions that they do not overlap in wing-length, which has in some instances led to mistakes in naming, as it was thought that forms differing so markedly in size could not belong to the same species.

Different species of swifts are affected by Gloger's and Bergmann's rules to a different extent. Thus in arid Damaraland, the African black swift is much paler, the alpine swift is somewhat paler and the African white-rumped swift is little if any paler, than their counterparts in South Africa. The same holds for other species elsewhere and also for the variations in size. These differences may be due to the extent to which each race is isolated from its neighbours, and to the length of time for which it has been in the area to which it is specially adapted. Both Gloger's and Bergmann's rules, it may be added, hold in mankind, in which the northern races tend to be fair and tall, and the tropical races dark and small, but there are exceptions, and in man the situation has been much confused by emigration and the wandering of tribes.

To determine the right names for the different forms of swifts, and their correct classification, was in some respects tiresome work, but it has involved far more than the arbitrary arrangement of a convenient catalogue. Moreover, when the correct names had been found, many seemingly trivial differences in colour and size were found to be adaptations to the needs of the birds. All studies of adaptation require, as an essential first step, a sound classification.

Chapter 16

THE BIRTH-RATE

At Oxford, most swifts start laying in the last week of May, and most young are in the nest between about 11 June and 8 August. As can be seen from Fig. 22, the longest and sunniest days of the year come near the beginning, and the warmest days near the end, of the swift's nestling period, which seems to be a compromise between the two. Since insects are most plentiful in the air in sunny and warm weather, it seems likely that the swift's nestling period

	MAY				JUNE				JULY					AUGUST			
	7	14	21	28	4	11	18	25	2	9	16	23	30	6	13	20	27
Most young swifts																	
Most sunshine (>6 hours)																	
Warmest (59°-60°F)																	
Longest days																	

Fig 22: Breeding season and climate

corresponds with the season when it can most easily find food for its young. The little that is known about the abundance of airborne insects supports this idea. It does not follow, of course, that a swift lays its eggs in late May because it is consciously aware that food will be most plentiful a month later, when its brood will need it. Probably, those swifts with an inherited tendency to lay their eggs in late May raise, on the average, more young than those which start earlier or later, and so they leave more young to carry on this tendency. It is not yet known how swifts recognise that late May has come, or, to speak more precisely, what factors in the environment (such as day-length or temperature) provide the signals which stimulate laying at the right season. Whatever factors may be responsible, the timing is accurate, for in ten years at Oxford the average date of laying has varied by only ten days, being slightly earlier with warmer and slightly later with colder weather in May.

Swifts normally lay two or three eggs in a clutch, and just as their date of laying seems adapted to feeding conditions after the young have hatched, so the number of eggs laid corresponds with the number of young that the parents can raise. In most years some of the nestlings have died through failure of the parents to bring them enough food, and the proportion dying has been larger in larger broods. A few broods (due to hatching failures) have consisted of a single nestling, and of these 94 per cent have been raised, while in broods of two 82 per cent, in broods of three 72 per cent and in the one brood of four 50 per cent, of the young have been raised. A few nestlings have fallen accidentally from the nest, but nearly all other deaths have been due to the inability of the two parents to bring enough food.

The losses from starvation vary greatly with the weather, an example of which is shown graphically in Fig. 23 for two broods, each of three young, one of which was raised in fine weather in 1949 (above) and the other in wet weather in 1947 (below). In 1949, all three of the young put on weight rapidly and flew successfully when 5½ to 6 weeks old. In 1947, on the other hand, one of the nestlings grew scarcely at all and died of starvation on the ninth day, another died before it was a month old, and only one of the three flew, several days later than normal and much below the usual weight.

The survival of nestling swifts in different years is summarised in Table 1, in which the years have been arranged in descending order of suitability for feeding, a convenient measure of the weather being given by the average maximum temperature during the six weeks

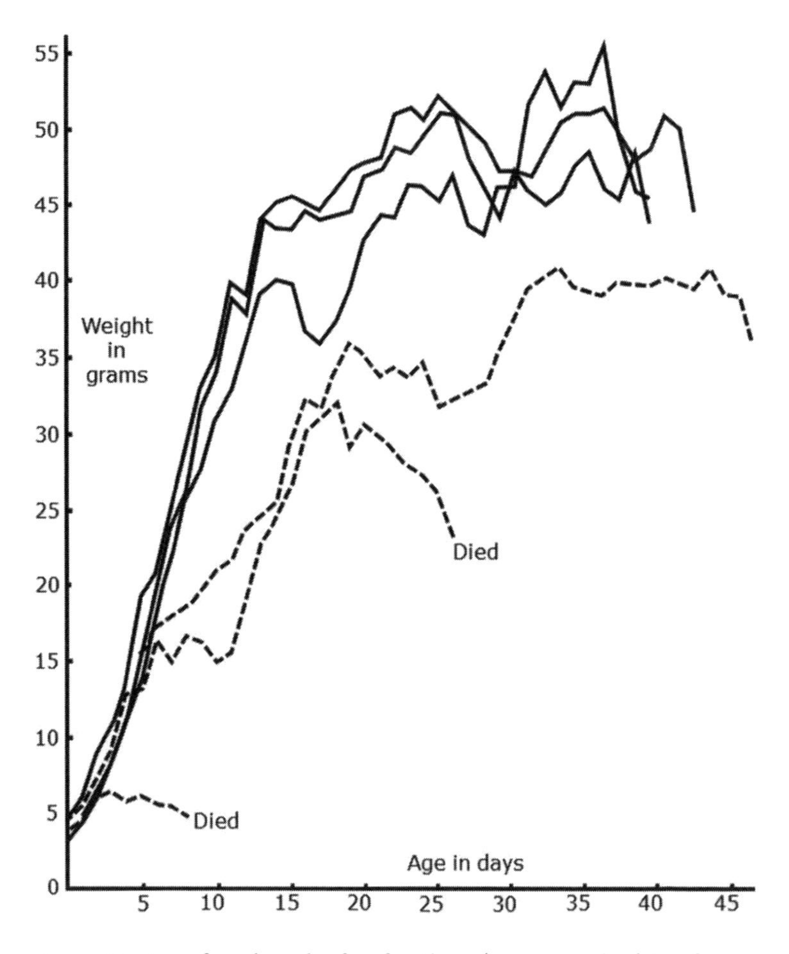

Figure 23: Fate of two broods of swifts. Above (continuous line) in a fine summer, below (dotted lines) in a wet summer

after the hatching of the first chick. On this basis, 1949 and 1955 have been the best seasons, and 1948 and 1954 the worst seasons, so far experienced.

TABLE I

NESTLING SURVIVAL

Year	*Average maximum temperature*	*Proportion of young raised*	
		Broods of 2	*Broods of 3*
1949	75°F	90%	100%
1955	73°F	90%	90%
1951	71°F	100%	90%
1952	70°F	100%	70%
1950	69°F	100%	70%
1953	69°F	100%	60%
1948	66°F	50%	50%
1954	66°F	30%	—

Notes: (i) The years are arranged in descending order of average maximum temperature, and where two figures are equal, the year with the higher rainfall is put second.

(ii) Percentages are given to the nearest 10 per cent, and are based on the following totals, reading from top to bottom: for broods of two, 26, 28, 20, 20, 28, 16, 16 and 30; for broods of three, 6, 9, 9, 27, 3,15 and 6.

(iii) In the one brood of four, in 1952, 30 per cent survived. Broods of one were raised successfully every year except for one out of seven in 1954 (and two accidental deaths in other years).

The influence of the weather on survival depends greatly on the size of the family. Every year the broods of one young received enough food, except for one which hatched during the worst spell of weather in 1954 and died of starvation after a fortnight. But the other six single young in 1954, and three in the almost equally bad summer of 1948, were raised successfully. (The only other deaths in broods of one were probably due to accident.) The young in broods of two were raised successfully except in the two worst years, 1948 and 1954, when respectively only a half and a third of them survived. (There were also a few accidental deaths in fine summers.) In broods of three, the parents raised all the young only in the extremely fine summer of 1949, though the one death in the fine summer of 1955 may have been accidental, and only one chick starved in 1951. In the two middling summers of 1950 and 1952 about two-thirds of the young were raised, and in 1953 (which was wetter than 1950) only 60 per cent survived. It is unfortunate that there were too few figures to give a reliable average for the bad summers of 1948 and 1954.

The most efficient size of family is, of course, that from which the greatest number (not the highest proportion) of young survive. Hence provided that slightly more than two out of every three young are raised, broods of three are more efficient than broods of two, even if the latter are wholly successful. For the uppermost five years in Table 1, three was therefore the most efficient clutch-size for the swift. But in poor weather in 1953, with a 60 per cent survival, an average of only 1.8 young was raised from each brood of three, as compared with 2.0 from each brood of two young. This suggests that in poor weather a clutch of two may be more efficient than a clutch of three, and one might expect the balance to shift more strongly in favour of broods of two in really bad summers. Actually, in 1954, the losses were so heavy even among broods of two that, on average, more young were raised from broods of one than two (0.9 young per brood of one as compared with 0.5 young per brood of two). In bad weather small families are more productive than large ones because, with more young to share the food, all are likely to be weakened when food is scarce. The effect of such competition, and its relaxation, can be seen in the lower brood in Fig. 23, in which the death of one nestling was followed by a marked rise in the weight of the remainder.

Only twice in ten years have we recorded a clutch of four eggs, and from the only brood of four, in 1952, two young were raised. Likewise in two broods each of four young in Switzerland, both in fine weather when all the young in broods of three were raised, only five out of eight young survived. More figures are needed, but it looks as though swifts cannot raise as many as four young and that, if they start with four, they end up with fewer than if they start with three. Hence the swift's clutch of three eggs has probably been evolved through natural selection because it corresponds with the most effective size of family. That, even so, some of the early layings consist of two eggs may be because of unfavourable weather at the time of laying; also in poor weather, families of two young give rise to more survivors than broods of three, so a clutch of two holds the advantage in some years.

In Switzerland, the summer is normally much finer and warmer than in England, and nestling swifts survive much better. Thus in the years 1947–54 inclusive (but omitting the bad summer of 1948), 99 per cent of the young in broods of three were raised successfully in Weitnauer's colony at Oltingen. At Oxford in the same period, only 71 per cent of the young in broods of three survived. Since in Switzerland swifts can nearly always raise three young, it is curious that all the pairs do not there lay three eggs; as in England, a clutch of two is not uncommon. In 1948, the one bad summer in Switzerland, the weather was much worse than in England, because the frequent rain was accompanied by severe cold. Out of thirty-four nestlings hatched in broods of two and three young in that year only six, or 18 per cent, were raised, a poorer rate of survival than any so far experienced at Oxford.

The survival of nestling swifts depends on the frequency with which the parents can feed them, to determine which we carried out a series of watches in the tower from 8 a.m. to 6 p.m., sometimes on fine and sometimes on wet days. The results are summarised in Table 2. In fine weather, broods of one were fed in the ten hours about

TABLE 2

NUMBER OF FEEDS IN TEN HOURS
(8 a.m.–6 p.m. GMT)

Brood	*Number of records*	*Average number of feeds*	
		per brood	*per young*
		IN FINE WEATHER	
1	21	8.9	8.9
2	59	14.9	7.4
3	18	19.3	6.4
		IN POOR WEATHER	
1	8	7.4	7.4
2	33	7.1	3.5
3	10	6.5	2.2

9 times, broods of two about 15 times and broods of three young about 19 times. This shows that, with more young to feed, the parents work harder, but the increase is not proportionate to the number of young. Thus dividing the number of visits by the number of young in the brood, each single nestling was fed on the average 9 times, each in a brood of two about 7½ times and each in a brood of three about 6½ times, in the ten hours. Some years earlier, R. E. Moreau and two African assistants obtained similar results from long watches outside the nests of the house swift, white-rumped swift and palm swift in Tanganyika, finding, as we did, that with a larger brood the parents brought food more often, but that nevertheless each nestling in a brood of two was fed less often than each single chick, and each

nestling in a brood of three was fed less often than each in a brood of two.

In bad weather at Oxford, the feeding rate was much reduced, the parents coming about 7 times in ten hours irrespective of whether the brood consisted of one, two or three young. Evidently in bad weather the parents are extended to their fullest to meet the demands of a single nestling. As a result, each nestling in a brood of two was fed half as often, and each in a brood of three one-third as often, as each single nestling. Comparison with the figures for good weather shows how bad weather affects the young chiefly in larger broods. Thus each single nestling was fed 7 times in bad weather as compared with 9 times in good weather, a small difference, but each nestling in broods of two was fed 3½ times in bad weather as compared with 7½ times in good weather, and each nestling in broods of three was fed only 2⅕ times in bad weather as compared with 6⅖ times in bad weather.

In Switzerland, as already remarked, nestling swifts put on weight more rapidly and have a higher rate of survival than in England. We had therefore expected that they would also be fed more often, but Weitnauer recorded a feeding rate similar to that in fine weather at Oxford. Instead, the parent swifts tend to bring larger meals in Switzerland.

The view that the normal clutch of the swift corresponds with the largest number of young that it can raise is supported by what happens in the alpine swift. In the large colony at the Jesuit church in Solothurn, mentioned in earlier chapters, H. Arn has recorded the number of eggs laid and young raised over many years. In the alpine swift, nearly all clutches laid at the normal season consist of three eggs, and the figures indicate that three is the most effective brood-size. Thus in broods of three 79 per cent of the young were raised, an average of 2.4 per brood, but in broods starting with four young only 55 per cent were raised, an average of 2.2 per brood. Hence although broods of four start with an additional nestling, their losses are so high that they give rise to fewer survivors than each brood of three. (The proportion of the young raised from broods of one and two was 97 per cent and 87 per cent respectively.)

A complication not so far mentioned, which applies to both common and alpine swifts, is that clutches laid later in the season tend to be smaller. At Oxford, of the clutches started in the first week of the laying period, up to 24 May, about two-thirds have consisted of three eggs and one-third of two eggs. During the following week, the last in May, one-quarter have consisted of three eggs and three-quarters of two eggs, while of those laid in June only two (out of forty) have been of three eggs. Similar variations occur in Switzerland. If clutch- size corresponds with the number of young that can be raised, this implies that swifts are more likely to succeed in raising three young earlier than later in the season. The catches in aerial tow-nets indicate that air-borne insects are more plentiful for the earlier than the later broods of the swift, and the available figures suggest (though they are not enough to prove) that late broods of the swift survive rather less well than the early ones. In the alpine swift also, late layings often consist of two instead of three eggs, and for this species the figures establish that the later broods survive less well than those at the normal season.

While the chief disadvantage of large as compared with small broods is the higher mortality among the nestlings, our measurements show that the young in larger broods also grow rather more slowly and stay for about two days longer in the nest, which may impose a somewhat greater strain on the parents. Further, when the young from larger broods leave the nest, they weigh rather less than those from smaller broods, which may reduce their chances of survival afterwards.

The number of eggs in the clutch varies in different types of swifts. Crested swifts (*Hemiprocne*) lay only one egg, most black swifts

(*Cypseloides*) one, most cave swiftlets (*Collocalia*) two, typical swifts (*Apus*) and Old World palm swifts (*Cypsiurus*) two or three, New World palm swifts (*Tachornis*) and scissor-tailed swifts (*Panyptila*) probably three, white-throated swifts (*Aeronautes*) and spine-tails (*Chaetura*) usually four or five, sometimes even six. This suggests that, as compared with our own species, crested and black swifts find it harder, while white-throated and spine-tailed swifts find it easier, to collect food for their young; but the point has not been studied.

The breeding season also varies markedly in swifts, even within the genus *Apus*. Thus at Oxford the common swift raises only one brood a year, and the length of the summer would not permit more. In the Mediterranean region, where warm weather lasts for longer, the very similar pallid swift raises two successive broods, the first clutches being laid in April, while the second broods do not leave until late September. In Tanganyika, where warm weather lasts longer still, the African white-rumped swift raises three successive broods each year, laying its next set of eggs only two or three days after the young of the previous brood have flown. In Zanzibar, again, where there are two rainy seasons in the year, the house swift has two breeding seasons, the first from September to January and the second in May. Hence in swifts, as in other birds, the breeding season is highly adaptable and is in general suited to the local conditions.

Little work has been done on the nesting success of other swifts. In a colony of the white-rumped swift in Tanganyika, 88 per cent of the eggs hatched and 86 per cent of the nestlings were raised, meaning that about three-quarters of the eggs laid gave rise to flying young. For comparison, at Oxford between 1948 and 1955, three-quarters of the eggs hatched and 86 per cent of the young hatching in broods of one and two were raised; but if broods of three are included, nestling survival was rather lower than in the white-rumped swift (in which there were no broods of three). Nesting success was decidedly lower in a colony of African palm swifts, in which two-thirds of the eggs hatched but only one-quarter of the nestlings survived, so that only one-sixth of the eggs laid gave rise to flying young. The palm swift does not, like most other swifts, nest in a hole but on a palm leaf, and nearly all the losses were due to eggs and young being taken by fiscal shrikes. Even those swifts with concealed nests are not always safe from enemies in Africa. Thus driver ants overran a colony of house swifts nesting under the eaves of a house in the British Cameroons, destroying twenty-five nests, most containing young, before the observer could divert the ants with kerosene.

In conclusion, it may be of interest to work out the increment in the swift population each year due to breeding, allowing for the losses of eggs and young. In the years 1948–55 inclusive at the tower, the average number of young raised annually by each breeding pair has been 1.3, varying from 0.6 in the bad summer of 1954 (when 25 pairs raised only 14 young), to 1.9 in 1952 (when 22 pairs raised 42 young). A higher figure is possible, since as yet there has been no year in which large clutches were combined with high survival. In Switzerland during the same period of years, the annual increment has been greater, in one year being as high as 2.6 young per pair, and averaging 1.8 young per pair per year. This means that each 100 pairs of swifts has raised, on average, 50 more young each year in Switzerland than in England. The increment is probably rather higher in the African white-rumped swift, since at the colony in Tanganyika mentioned earlier, most pairs laid three successive clutches of two eggs each, while three-quarters of the eggs laid gave rise to flying young, so that each pair probably raised more than four young a year. In the colony of palm swifts, on the other hand, 19 pairs raised only 6 young, or 0.3 per pair.

Chapter 17

DEATH AND ITS CAUSES

A popular bird-book of the nineteen-twenties was entitled *How Birds Live*. There is now need for a companion volume on *How Birds Die*, a problem of immense difficulty, as naturalists find only a small fraction of the dead birds. Partly for this reason, many bird-watchers have refused to believe the high figures that have been given for the death-rates of wild birds, though the ways in which these estimates have been made seem sound and they fit with the known birthrates to give 'balanced' populations. Proof that naturalists see only a small fraction of the deaths that occur is shown by the fact that of every hundred ringed song-birds only one or two are ever found again and reported. The rest die undetected. In swifts, which migrate and range over a great area, the proportion found dead is even smaller. Thus in the twenty years 1931–50, of 669 swifts ringed in Britain as nestlings only 4, and of 560 ringed as adults only 3, have later been found dead away from where they were ringed. The Swiss observer Weitnauer has studied marked swifts for long enough to permit a rough estimate to be made of their survival. Eighteen individuals caught and ringed in his nesting boxes returned, on the average, for another 5½ years. If, as is likely, all that failed to return had died, this figure represents the average further life of an adult swift. It corresponds with an annual loss of one-sixth of the birds.

Another estimate has been made from the records of 650 breeding swifts caught, marked and recaptured between the years 1930 and 1947 at Hasselfors in Sweden. By no means all of those known to be alive were recaptured each year, but allowance could be made for this. Also, losses were much heavier in the first year after ringing than any other, probably because some of the birds deserted when first captured and moved elsewhere. With the error from this last cause allowed for, the average further survival was four and a half years, just under one-fifth of the birds being lost each year. The losses may well have included some later desertions, so the death-rate was probably lower than the figures suggest. The oldest adult survived at least another seventeen years after it was first caught.

The most satisfactory estimate yet made for the survival of swifts is based on the ringing recoveries of 353 adult alpine swifts caught, marked, and in many cases recaptured in later years, at the Jesuit church in Solothurn over the period 1920 to 1950. Here, as at Hasselfors, allowance has to be made for the fact that not all the birds known to be alive were caught each year, but there was no reason to think that any deserted the colony as a result of capture. About one-sixth of the adults were lost each year, giving an average further life of five years.

Hence estimates of the death of swifts from three different sources are very similar, particularly since the Swedish losses probably included some desertions. It may seem shocking that as many as one-sixth of the adult swifts die each year; certainly, we would be horrified at so high a loss among our friends in the prime of life. Actually, swifts have a lower mortality than any other wild birds so far studied on a large scale in Europe or North America. Only the yellow-eyed penguin in New Zealand has been found to live longer, with an average further life of over seven years, one-eighth of the adults dying each year. On the other hand, in many of our familiar song-birds about half of the adults die each year, and the same is true of various game-birds, ducks and doves, while in certain herons, wading birds, gulls and terns one-third of the adults die each year. Many song-birds can expect to live for less than one and a half years.

This high annual loss is a result of the high birth-rate. Over the course of a few years, most bird populations show no overall change in numbers, which means that birth-rate and death-rate are equal.

There is a popular idea that this is because the birth-rate is somehow adjusted to the death-rate, long-lived birds producing only a few young because only a few are needed to replace losses. The true explanation is the other way round. The birth-rate, as discussed in the last chapter, probably corresponds with the highest number of young that the parents can normally raise. But if a population is held in check, whether by starvation or by enemies, then a higher birth-rate cannot normally cause an increase in numbers, but merely means that more individuals die each year.

The figures given in this and the previous chapter can be combined to work out the turnover in a population of swifts. A death-rate of one-sixth means that 17 out of each 100 adults die each year. When numbers remain steady, these deaths are balanced by 17 newcomers breeding for the first time. Swifts first breed when two years old. Assuming, as is not unreasonable, that their death-rate in the second year is similar to that of the adults, then 17 two-year-olds have survived from 20 yearlings. From how many fledglings have these 20 yearlings come? At Oxford each pair of breeding swifts has raised, on average, 1.3 young each year, which means that each 100 adults (50 pairs) raises 65 young. In a stable population, these have been reduced to 20 yearlings by the following summer, a loss of about 70 per cent. Hence the mortality in the first year is about four times as great as in any later year. This seems not unreasonable, as the young bird is inexperienced. In other species also, juveniles have heavier losses than adults, though the difference is not so great as in swifts. Actually, if we had used the Swiss figure of 1.8 young raised per pair, instead of the English one, the mortality in the first year would have been higher, 78 per cent.

How do swifts die? The answer is largely unknown, and since as already mentioned fewer than one in every hundred swifts is found dead, the few that are found are unlikely to be typical. A huge German compilation of the prey taken by European hawks and owls shows that swifts are rare victims. Thus of 4,261 birds of all kinds caught by sparrowhawks only 51, just over 1 per cent, were swifts. Similarly swifts made up only 1½ per cent of the recorded prey of the peregrine falcon and 2¼ per cent of the prey of the hobby. The hobby has sometimes been considered a serious enemy of swifts, but in fact it takes chiefly swallows and skylarks, and even the percentage of swifts taken is misleadingly high, since hobbies also catch many insects, including dragonflies, which do not leave bony remains and so were omitted from the analysis. Swifts feature extremely rarely in the diet of other European birds, though kestrel, tawny owl and barn owl take them occasionally.

In a spell of cold and wet weather in the third week of August one year, a German observer found that of 67 birds taken by hobbies as many as 26 (about 40 per cent) were swifts. Hence swifts were nearly twenty times as common as usual in the prey, obviously because they had been weakened by undernourishment and hence were much easier than usual to catch. But under these circumstances starvation, not predation, should be reckoned as the predisposing cause of death.

Likewise in the German town of Konstanz in the extremely cold and wet summer of 1948, the remains of many swifts were found in each of three kestrels' nests and others were obtained from the pellets of barn owls. At this time, it will be remembered, swifts clung to walls half-torpid, while some fell dead to the ground, so that neither kestrel nor barn owl need have caught them in the air. A kestrel has, however, been recorded diving into a swarm of swifts and capturing one in flight. Also, one evening when I was inside the tower, I heard a sharp knock on a box just after a swift had entered it for the night, and my wife watching outside saw a tawny owl strike at the passing bird, though it was much too slow to catch it.

Swifts sometimes gather behind a flying hobby, sparrow-hawk, or even the harmless buzzard. A swift in good health can probably outfly a bird of prey in level flight, so that it is safe for it to follow behind. Probably, both hobby and sparrowhawk catch swifts chiefly by coming among them at full speed before the swifts are aware of them. Even then, matters are not always easy for the hawk. Thus an observer saw a hobby in laboured flight which eventually landed. He caught it and found that it had taken a swift, but the swift had its claws fixed in the hobby's thigh while its shoulder feathers had become entangled with the hobby's cheek feathers, thus forcing its head down so that it could not fly properly. The birds were ringed and released, both apparently unhurt.

Wild mammals are not serious enemies of swifts. Rats or weasels have sometimes found their way under the roofs where swifts are nesting and have killed nestlings, while Gilbert White recorded a cat catching an adult swift as it swooped to enter its nest under a very low roof. One other mammal also destroys swifts at times. We formerly supposed that we were the first to construct a tower where swifts might dwell, but how rarely is one the first in any human achievement. In Tuscany, towers have long been prepared as nesting places for swifts, though for a meaner purpose than ours. In one instance a special tower was built on a steep rock rising out of a stream, with holes in the walls for the swifts to enter by and ledges inside for them to nest upon. More usually, the top of an existing tower or the highest wall of a house was adapted in a similar way. By this means, at least until the end of the nineteenth century, nestling swifts were regularly taken for the table, their flesh being described as delicious (a point that we have not checked). One youngster was usually left in each nest and the adults were not taken, as their flesh is tough.

Italians used also to take adult swifts. Thus when Charles Waterton visited Rome in 1817, he saw boys catching swifts and house martins by attaching a small feather to a silk line with a noose behind and floating it above the houses, the bird taking the feather for its nest and entangling itself in the noose. The bodies were then sold in the bird-market. 'This ornithological amusement is often carried out in the street of the Propaganda during the months of May and June.' In the *Compleat Angler*, likewise, Izaak Walton noted that in his day swifts were often taken with rod and line in Italy, and they have sometimes been caught accidentally at the present time by anglers on English rivers.

The swift also has enemies smaller than itself, though most of them cause irritation rather than serious harm. The most striking of its parasites is the flightless louse-fly *Crataerina* (Fig. 24, Plate 18), which is one-third of an inch long, with a triangular, bottle-green body and long thin legs with efficient gripping claws. It runs actively forwards or sideways in the feathers of the swift and sucks its blood, being particularly attached to the nestlings. It is about three times the size of a human body-louse, and since a swift is only six inches instead of six feet in length, it is as if a man had body-lice four inches long rushing about in his clothes and hiding skilfully in the creases when he tried to catch them. *Crataerina* (see Plate 18) looks sinister, particularly when, after we have been handling young swifts and are back in company, one shyly sidles from under the collar and down the jacket. Once I was bitten, but normally they do not hurt a man and soon leave him.

It has been found that each *Crataerina* takes a meal once in five days, then feeding for about ten minutes and taking about 25 milligrams of blood. With a swift weighing 42 grams, the proportionate amount of blood taken is about one-quarter of that given by a man in the voluntary blood transfusion service, but a man is given six weeks in which to recover, whereas the *Crataerina* will seek another meal in five days, and we have found up to twelve of them on a single nestling swift. Even so, they may cause no appreciable harm to well-nourished young, though in a bad summer when the nestlings are short of food, *Crataerina* perhaps increases the losses.

The life-history is beautifully adapted to that of the host. Like other hippoboscid flies, *Crataerina* does not lay eggs. Instead, a

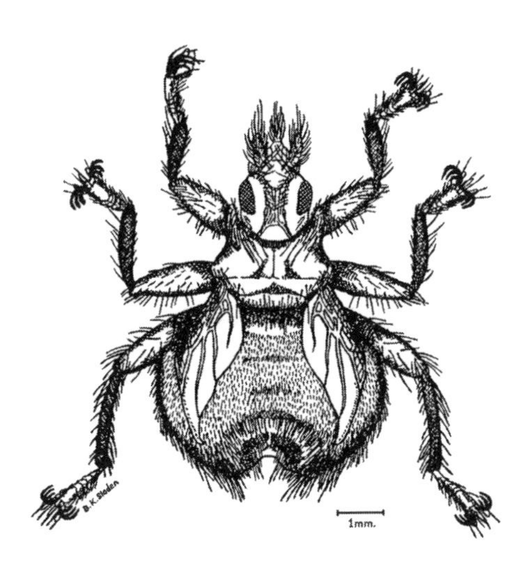

Fig 24: *Crataerina*

single larva is produced at a time, which pupates at once. Hence all the feeding is done by the adult, the larva obtaining the nourishment for its growth while inside the female. The pupa is a small brown capsule, looking rather like a vitamin pill. The adult insects die when the swifts depart in the autumn and only the pupae, which lie in the dust and crannies close to the swift's nest, survive the winter. The adults emerge from them in June at about the same date as the nestling swifts hatch. It has been shown that this neat timing is due not, as might have been thought, to the warming of the nest by the brooding swift, since many of the pupae are well outside the nest, but to the general rise in temperature in the spring. When *Crataerina* pupae were transferred in winter from a nesting place to an incubator set at the temperature normal for June, the adults emerged from them two months earlier than usual. An unsolved problem is how these flightless insects find their way into new nesting places. As already mentioned, swifts (probably those seeking breeding places) sometimes visit the holes of other pairs, and at times they move from one colony to another. This seems the only way in which the parasites could be transported.

Once in early June in Hertfordshire, a flying swift fell suddenly to the ground and died. Twenty louse-flies were collected on it and several more escaped. The Swiss observer Weitnauer has likewise reported five separate instances in which a heavily infested swift came to the ground, four of the birds dying and the fifth being very weak. It therefore seems that *Crataerina* is capable of killing an adult swift, but this is perhaps abnormal, as its life-cycle seems timed to fit with the appearance of the nestling birds. Further, those *Crataerina* which kill an adult swift away from its nesting hole must quickly die themselves, while their offspring have no chance of finding another swift in the following year.

Swifts, like other birds, have feather-lice (Mallophaga), while mites and fleas occur in their nests, but all of these seem harmless. Their feather-lice are classified in two genera, *Dennyus* and *Eureum*, which are peculiar to swifts, and swifts lack another group, the Ischocera, which are found in all other birds. The Mallophaga provide valuable clues to the affinities of the main groups of birds, and this peculiarity suggests that the order of swifts (Apodi) evolved from other birds far back in the past. A fossil wing-bone attributed to a swift has been found in deposits of Eocene age, which is the time when the main orders of birds are thought to have originated.

The other species of swifts, like our own, have few natural enemies. Cave swiftlets (*Collocalia*) are preyed on by a variety of hawks which wait for their mass entry into the nesting caves at dusk, but the proportion taken in this way is presumably small. Occasionally, both black swifts (*Cypseloides*) and spine-tailed swifts (*Chaetura*) have been caught by falcons. A surprising record is of a chimney swift found inside a three-pound bass, the fish presumably taking the bird as it skimmed low over the water. But in these other species, as in our own, natural enemies probably account for only a small proportion of the annual losses.

The basic problem presented by the death of swifts is how the numbers of the species remain more or less constant from year to year. This must mean that when, for any reason, swifts have become more numerous than usual the death-rate rises, so that they soon decrease again, and conversely that when swifts have for any reason become scarcer than usual the death-rate falls, so that they soon increase again. The problem can scarcely be solved in Britain, for the critical mortality must affect the adult swifts, and although about one-sixth of the adults die each year, we have rarely recorded a death at the tower.

The main losses evidently occur after the swifts have left us, either in their winter quarters or on their way there and back. Food shortage seems likely to be the main cause of death, since swifts have few natural enemies and these perhaps catch them chiefly when they are undernourished. In Africa, swifts wander over great distances, appearing when flying ants or termites are abundant and then moving elsewhere. It is extremely hard to visualise their being short of food during part of their stay, yet I do not see any other way in which their numbers can be kept in check. There is the further problem that in Switzerland, as already mentioned, breeding is more successful than in England (by as much as half young per pair per year). With a stable population, this means either that the Swiss birds have a higher death-rate than the English ones, or that swifts are continually leaving Switzerland for less favoured localities. There is no evidence for the latter movement, and it seems unlikely that it could be happening on the large scale required. On the other hand, why should Swiss swifts fare less well than English ones in Africa? Perhaps, like various other migrants, those from different breeding areas have different winter homes, but even so the problem seems extremely difficult. It is unlikely that it will ever be solved. The same problem has not yet been solved, and will be hard to solve, even in a resident ground-feeding bird. In the wide-ranging aerial swift it is hard to suggest how one could even start seeking for an answer.

Chapter 18

THE MEANING OF ADAPTATION

The swift is a wonderful example of adaptation. In this it resembles all other forms of life, but because it is an extremist the adaptations are more striking. It can move through the air faster than any other bird of its size, helped by its long narrow wings and associated bones and muscles. Because it has long narrow wings, it is slow in taking off, but it selects a nesting place with a clear drop in front. Its nesting hole is often cramped but it can mate on the wing. It utilises a food supply on which no other British bird depends and when bad weather makes air-borne insects scarce, it can fly round a thunder-shower or travel for several hundred miles to avoid a wide belt of rain. In cold weather adult swifts may come together in clumps, while their eggs and young can survive long periods of chilling. The nestling can also survive for a long time with little or no food and saves energy by growing more slowly and cooling down at night. At the end of the summer, when swifts could no longer survive here, they fly for thousands of miles, the young before the parents, and reach southern Africa.

The swift, then, is highly specialised for rapid flight, while the disadvantages that this might have brought are overcome by other remarkable adaptations. At the same time, it should be kept in mind that what to us seem more ordinary birds are equally remarkable. Thus a robin cannot fly so fast, but it can turn or stop with greater ease and precision, which is important for its life among bushes. Again, while a nestling swift can grow more slowly when food is short, thus saving energy, it is equally remarkable that under similar conditions all the members of a brood of robins grow their feathers at a similar rate, so that they can leave the nest on the same day. Each of these species is fitted to its own way of life, and when all is said, Britain supports many more robins than swifts. After marvelling at swifts and eagles, gazelles and elephants, let us remember that the world is inhabited chiefly by sparrows and rats.

All animals are adapted, and the more that is discovered about them the more intricate and efficient their adaptations are found to be. That each seems well designed for its place in nature gave, it was formerly supposed, sure evidence for a Designer. This argument, of course, was put forward long after men came to believe in God, it was not the reason for belief, and it was demolished, at least in the form made popular by Paley, by the theory of evolution. The time has come to leave the swifts in their tower, but as we leave, let us pause once more on the floor below, to see what has since happened to the dispute between Wilberforce and Huxley. After a bitter struggle, the occurrence of evolution came to be accepted as proved, even for man. Darwinism further showed, or seemed to show, that the wonderful adaptations of animals were the result of a wholly natural process of selection acting on hereditary elements. The nature of inheritance was not at that time known, but this knowledge came at the start of the next century, with the rediscovery of Mendel's laws; and the inference commonly drawn from these findings was that animal adaptations, and man himself, came into existence through Blind Chance.

This led to the second phase of the attack on Darwinism, one not yet over, which has been directed against natural selection. The scientific objection was voiced by Samuel Butler: 'Shall we maintain that the eagle's eye was formed little by little by a series of accidental variations, each one of which was thrown for, as it were, by dice? We shall most of us feel that there must have been a little cheating somewhere with these accidental variations before the eagle could have become so great a winner.' The argument has been repeated, with different examples but in essence unchanged, up to the present day. Secondly there was a strong emotional dislike for a process which appeared 'to modify all things by blindly starving and murdering in the universal struggle for hogwash', as Bernard Shaw put it later. Thirdly, natural selection seemed to leave no place for human values. To quote Bernard Shaw again, 'there is a hideous fatalism about it, a ghastly and damnable reduction of beauty and intelligence, of strength and purpose, of honour and aspiration'.

Most of those who have rejected natural selection as the means of evolution have turned instead to some form of internal striving, a conscious or unconscious willing of the animal, a Universal Mind or Life Force. This has been particularly urged by those who, though atheists, have felt that the world exhibits purpose, and also by some Christians who have supposed natural selection to be incompatible with the higher nature of man. The view has never been popular with zoologists, who by detailed observations and experiments, helped in modern times by statistical analysis, have shown beyond reasonable doubt that natural selection occurs, that it is immensely powerful and, by inference, that it has been the main agent of evolution.

On one point Samuel Butler was right, that there must have been some cheating with the accidental variations. Purely random changes in the hereditary make-up could not create an adapted organism, but to introduce a Life Force to do the cheating does not solve the problem. No one has seen a Life Force or measured its effects, no one can say of what it is made or how it originates or is passed on. It is just a myth, a myth without divine sanction, and there is no better evidence for it than for the old gods which moved the waters or flashed the lightning.

The mistake arose through thinking that the direction of evolution is primarily set by mutations in the hereditary characters of the animal. Mutations indeed occur at random and many are harmful, only a few are helpful. But what modern zoologists, and in particular Sir Ronald Fisher, have shown is that natural selection is much more than the 'sieve to which it was at one time likened. Each mutation has many effects on the body and each part of the body is influenced by many hereditary factors. Natural selection works on combinations of factors, which means that it can work to create new combinations, and hence new and improved adaptations, far more quickly and efficiently than was once thought. It is natural selection, in fact, which provides the 'cheating' which Butler rightly saw was needed. Furthermore, since mutations often recur and are usually harmful, they would, without some corrective, soon cause later generations to be less well adapted. Natural selection is essential not only for promoting evolutionary change but also for conserving what has been evolved.

Natural selection works because in every kind of animal most individuals die before they have produced offspring, or at least before they have produced as many as they might. Thus about two-thirds of the young swifts which leave the nest die before they reach maturity, in most birds the proportion is higher, and in most other animals far higher still. The process seems wasteful and cruel, so that to many it seems terrifying, if not incredible, that man was evolved by its means. Earnest persons have therefore sought in nature for some mitigation, but the danger of a search inspired by such motives is that the evidence may be selected. The only true attitude is complete submission to the facts and, in the present connection, the immense annual destruction of wild creatures cannot be doubted. It is one part of the bewildering problem of pain, but as Dr Pusey commented, during the controversy over Darwinism, 'What are we that we should object to any mode of creation as unbefitting our Creator?'

It is wrong, also, to suppose that through natural selection evolution proceeds by 'blind chance'. For the effects of natural selection are orderly and consistent, not random. To take only one example, the Galapagos archipelago in the Pacific has been so isolated that most of the usual land-birds have not established themselves there. One bird, the Geospiza, has done so, and in the absence of other forms it has evolved into several finch-like seed-eating species, a parrot-like fruit-eater, tit-like and warbler-like insect-eaters and a woodpecker-like tree-climber. In short, evolution has resulted in forms very similar to those elsewhere. The many instances of convergent resemblance in animals cannot be ascribed to chance.

But even if evolution proceeds not by chance but by natural laws, are not the laws blind? That is not a question which the zoologist, from his special knowledge, is qualified to answer. To some people the existence of a universe governed by natural laws provides the plainest evidence for a Lawgiver, but to others it provides the plainest evidence for the reverse. Both theists and atheists have claimed support for their view from nature, but both use arguments which are in essence outside science and which carry little if any weight with their opponents.

The most serious argument against natural selection is that it seems to do away with values, with truth, goodness and beauty. Various agnostic biologists have sought to show how man's idea of good might have been evolved by natural selection, but the real difficulty is not so much its origin but why, if it has been evolved, any value need be attached to it. So far, at least, such attempts have given no sure guide for distinguishing and no valid reason for following what is good, and there are many who hold that, by their nature, all attempts along such lines must fail.

The dilemma was already clear to Darwin: 'With me the horrid doubt always arises, whether the convictions of man's mind, which has been developed from the mind of lower animals, are of any value or at all trustworthy.' But if man evolved wholly by natural means, as many scientists profess, Darwin's 'horrid doubt' would seem a certainty, and abstract truth, including the theory of evolution, can hardly be trusted. The issue, to repeat, is not so much whether a man could evolve moral feelings or intellectual discernment, but whether, if they have been evolved, his ideas of goodness and truth can be trusted. On the other hand, if truth and goodness have more than an arbitrary value, and most of us act as if they had, this might mean that they are related to what is outside man and nature, to the supernatural.

Hence while many of the side-issues in the dispute between Wilberforce and Huxley have been settled, the basic problem remains unresolved. Nor is it merely a point of academic philosophy, but one which vitally affects our lives and actions now. Many reject the supernatural, and either accept goodness as inexplicable, in which case there is no strong reason for following it, or, if they press the point, suppose man's beliefs and actions wholly determined, his free-will an illusion and truth and goodness of no ultimate significance, which seems in theory to be self-destructive and in practice to lead to evil acts. But this unpleasant consequence gives no good ground for rejecting natural selection, as various humanists have done. Could, then, man have evolved from the beasts by natural selection and yet apprehend truth and goodness through a supernatural gift? On this view there is a great gap in knowledge, but one which appears to involve no greater intellectual difficulty than those implicit, and often overlooked, in the alternative theories of man's nature. This is a grave matter with which to end a bird-book, but the tower where our swifts live was built by those who held that the study of nature should lead us, through a truer understanding, to a fuller worship of the Creator, and the times urgently require us to search out the basis of our lives.

Chapter 19

SWIFTS IN A TOWER – SIXTY-TWO YEARS ON

Dedicated to Roy Overall and George Candelin, successive 'Keepers of the Swifts', without whom there would have been no further study on the Swifts in the Tower.

My father, David Lack, published his landmark book, *The Life of the Robin*, in 1943. It was the fruit of a study he had done while working as a school teacher at Dartington Hall in the 1930s, having been fascinated by birds from an early age. This book described our most familiar and most loved bird, resident in these islands and usually tame. One would think that, if any bird was known well, it would be the robin, but he found all sorts of details that were not known, simply by observing the birds and marking individuals with colour rings. This was a revelation. He realised that almost any natural history observations of the kind he had grown up enjoying could, if done systematically, produce new results and ideas. Others realised this too, and it ushered in a golden age of natural history observations spawning seminal studies on other birds as well as insects and plants and the famous New Naturalist book series from Collins (later HarperCollins). Throughout his life it was natural history and patient observation of wild birds that excited him and was the basis for his entire career.

It was immediately after the war, in 1945, that he was appointed director of the Edward Grey Institute of Field Ornithology in Oxford and had the opportunity to study birds professionally. He had to abandon robins as their nests were too hard to find in quantity so started studying the tits in Wytham Wood, especially the great tit. But he could also see swifts from his office window in summer, and they fascinated him. Here was a bird that was about as different from a robin, or from any other common urban bird, as it could possibly be. Intensely aerial, migratory, staying here only for around three months a year, not closely related to any other group of birds and living in roofs. My father, like all of us, was in awe of the swifts' flying ability. He realised that, as many observations on robins were new, if only he could reach the swifts at their nests, everything he saw would be new. The possibility of studying swifts was irresistible and he conveys that sense of excitement in this book.

Since the publication of *Swifts in a Tower* in 1956 much more work has been done on swifts, some using the kind of observational methods that he enjoyed, but much of it going beyond that with some experimental manipulations and the use of new technology. The place of simple natural history observations in scientific study has, to a large extent, been taken over by these. The result is that we have many new details about the life of this extraordinary bird. I feel confident that my father would have been fascinated indeed by all the new insights about swifts that I outline here, although I am not sure that he would have wanted to work on them in what has inevitably become a more detached way. Direct human observation and involvement with the birds' lives was his style; the new technology, for all its advances, inevitably leads to a greater distance between observer and observed.

THE TOWER

In 1956 I was only two years old and the study of the swifts in the museum tower had been going for eight years, since 1948. Although the book was republished in 1973, shortly after my father's death, this was a reprint and it was not updated.[1] It has long been out of print.

[1] Curiously, the reprint took A. M. Hughes' lovely, but totally inappropriate, drawing of the Alpine Swift (p.170) as the central part of the cover. Clearly the publisher had not read the book, nor consulted my father.

But study of the swifts in the tower has continued almost unbroken since my father started it, first under his eye and subsequently under his successor, Professor Chris Perrins, and his colleagues. This is mainly thanks to the sterling persistence of Roy Overall (see Plate 19), an amateur ornithologist in the best of senses, who has climbed the tower more often than any other person. Every April, before the swifts arrived, he would check that the boxes were secure, and clean out some of the debris, especially on the glass backs of the boxes to enable clear observation from inside the tower. He has monitored all the significant moments in the life of the tower's swifts from arrival times, numbers of successful nests, timing of egg laying, hatching and fledging and has ringed the young birds and some adults. He did this every year between 1962 and 2010, meaning that, with an average of 25 visits per summer, he must have climbed the tower more than 1,200 times over forty-nine years. George Candelin joined Roy and shared the job from 1998 until 2010. From 2011–2013 the Edward Grey Institute assumed the task of monitoring the swifts, falling into the hands of Dr Sandra Bouwhuis in 2011 and Dr Andrew Gosler in 2012 and 2013. Since 2014 George Candelin has continued single-handedly, having been given the delightfully archaic title of 'Keeper of the Swifts'. The result is that we have one of the longest studies of its kind on any bird, rivalling that of the great tits in Wytham Wood.

There have been several changes to the tower since 1956. The swifts nested in what had been designed as ventilation holes in the original roof. The biggest single change for the swifts came in the refurbishment of the roof in the winter of 1965–66. Improved insulation was put in and the opportunity was taken to make the cowls slightly larger and replace the single swift nesting boxes behind each cowl with twin boxes, immediately doubling the number of boxes from 40 to 80. Then a further 67 swift boxes were added just under the eaves at the base of the sloping roof, bringing the total number of boxes to 147. A new floor and fixed ladders were added at the same time. The temperature extremes described at the end of Chapter 2 were undoubtedly lessened as a result, though it could still be hot. In the famously hot dry summer of 1976, when Derek Bromhall and his colleagues were filming the swifts in the tower, some of the nestlings crawled to the nest entrance to cool and then fell out to die on the baking tiles below. In any hot spell in the summer chicks can overheat and die, though cold wet summers are usually the greater danger because of food scarcity. Derek Bromhall's film, *Devil Birds*, was shown as part of the 'Survival' series on Anglia TV in 1980 in conjunction with publication of his book of the same title.

In 2010 there was considerable disturbance to the tower, starting with scaffolding on the west face, although this was removed in late April. Subsequently that summer there were evening performances with light shows on the lawn in front of the west face, followed by a large exhibition called the 'Skeleton Forest' erected using cranes. These may have disturbed the birds, although many swifts nest in buildings with active use, but it was that year too that a sparrowhawk struck (see later under *Death and its Causes*) in a summer of poor weather. The combination led to many pairs failing. This was immediately followed by two more summers with cool, wet weather, and in 2011 there was an attempt to use tiny loggers to track activities of a few of the birds and this led to rather greater disturbance that may have led to a few desertions (see *Numbers*). The swifts had had a difficult few years.

A further change, for the human observers, has been access to the tower. Ascent from the first floor of the museum still starts with the narrow, spiral stone staircase rising clockwise to the floor of the tower but the 30-foot wooden ladder leading up from there has gone. This bowed and bounced like a suspension bridge and could be challenging for anyone of a nervous disposition. In its place there is a solid metal spiral staircase, this time ascending anti-clockwise – so giving the sword-arm advantage to the attacker! Care is still needed, naturally, as it is always kept dark during the breeding season – there are fluorescent lights but these are locked off during the breeding season. Swifts are very sensitive to light at the nest and the glass backing of each nest is covered with a dark cloth; Roy and George always wore dark clothing.

As in all walks of life 'Health and Safety' legislation has inevitably intruded and, despite it being much safer than before, access is more restricted and controlled than ever. As we all know, this legislation has been taken to extremes, and in 2009, emergency floodlights were installed, to come on in a power failure or if testing generators. If these ever came on they would disturb the swifts, and some at least would almost certainly desert. Roy Overall successfully saw to their immediate removal and replacement with red-covered lenses as they would have been contravening the Wildlife and Countryside Act of 1981 – mutually contradictory pieces of legislation may become an increasing feature associated with health and safety.

Video links were installed, thanks to the generosity of Jocelyn Allard, on two of the swift boxes in 1996 that gave live views of the nests to a monitor in the main court of the museum. *The Museum Swifts*, which I wrote with Roy Overall, was published by the museum in 2002 to go alongside the video link. In 2004 four new video cameras replaced these, so four nests could be followed at once and transmitted both to the museum court (Plate 19) and onto the Internet via the museum website where an annual swift diary is published and updated weekly. All visitors to the museum can now see, in the light, comfortable, but crowded and noisy, surroundings of the museum's court what is going on more than 30 metres above their heads, in the dark, somewhat claustrophobic quiet of the tower itself. Indeed anyone around the world can access these links to watch and read about what is happening to the swifts, live, in their most intimate surroundings. In the museum, headphones can allow visitors to listen to the swifts too.

CLASSIFICATION (SEE APPENDIX TO CHAPTER 5 AND 'THE RACES OF SWIFTS' CHAPTER 15)

Swifts form a well-defined family or, more usually, two families with the South-East Asian treeswifts separated from the other swifts. They have no close relatives but have long been placed in the same order as hummingbirds, Apodiformes, on anatomical grounds, despite the obvious differences in appearance. The most similar birds in general appearance, the swallows and martins, are in an entirely different order, the Passeriformes or passerines, related to the thrushes, tits and other familiar garden birds, and look similar owing to a similar lifestyle catching aerial insects. There have been considerable changes in bird classification since 1956. This has come about for two main reasons: firstly the steadily increasing ability to study birds' biochemistry, starting with egg-white protein and enzyme differences and, since about 2000, to detect differences in DNA directly. Secondly, and taking DNA studies into account, there have been changing ideas about how we define a species.

DNA study has overturned much of what we thought about the broader relationships of bird orders and families and a new classification is only beginning to settle into mainstream publications. DNA study has confirmed the link between swifts and hummingbirds but these separated from the curious owlet-nightjars, eleven living species confined to Australasia, around 54 million years ago at the start of the Eocene. The owlet-nightjars are similar in morphology and habits to the true nightjars, potoos, frogmouths and oilbird that all separated slightly earlier on the same evolutionary branch (known as a clade by taxonomists). On current evidence, a true evolutionary classification would include all of these in a single disparate order, Caprimulgiformes (Fig. 25) or separate into five orders. It is noteworthy that the poor-will, a North American nightjar, is the only bird known to hibernate and that short-term torpor is known in some other nightjars, frogmouths and several hummingbirds as well

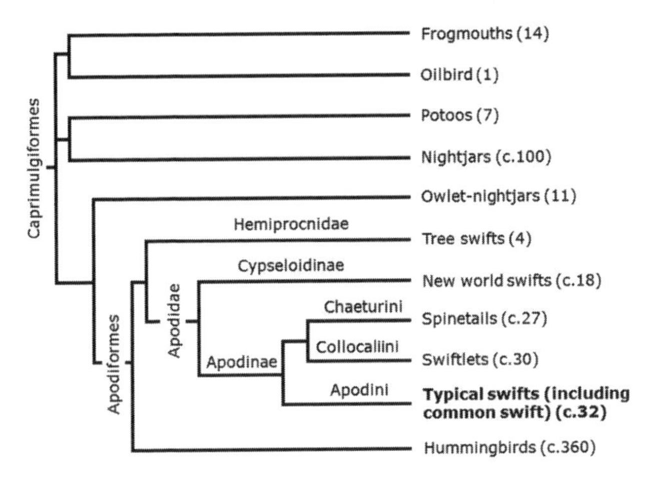

Fig 25: Family tree of swifts and their relatives

as young swifts. A few other birds can become torpid but it appears to be a widespread feature on this clade.

The 'Biological species concept' has held sway since Ernst Mayr first formulated it in the 1940s and is still taught in schools, but has been largely superseded in current classifications by a 'Phylogenetic' or 'Evolutionary' species concept. One of the problems with the Biological Species Concept is that it is defined as individuals that can actually or potentially reproduce to produce fertile offspring. Hybridisation has been shown to be much more common than earlier realised and the definition was obscuring important differences between different populations or subspecies. The more recent species concepts define species more narrowly, based on 'independent evolutionary trajectory' even if hybridisation is possible. The result is that some birds formerly described as geographical subspecies have been elevated to species status, as have some 'cryptic' species, those looking almost exactly alike, that have been shown to be isolated from each other on DNA grounds. The result is a greater number of species overall – taxonomists are currently in a phase of 'splitting', similar to that of the 19th century, as against the 'lumping' of the mid-late 20th century. Swifts have been much affected by this. In the Appendix to Chapter 5 my father described 65 species of typical swift in eight genera and three 'crested swifts' (the treeswifts) in one genus in the closely related family Hemiprocnidae. Today the number is just over 100 species of typical swift in 19 genera, though the status of some is disputed, and four treeswifts (still one genus).

The status of the subfamilies has changed too, with the Apodinae including all except the 13 or 14 species of Cypseloidinae (formerly just the genus *Cypseloides*, now two genera), all confined to the Americas (Fig. 25). The distinguishing feature of the Cypseloidinae is that they are the only swifts that do not use saliva to build their nests and are thought to be the most primitive group. The other feature that distinguishes them from our swift is the arrangement of the toes. They have feet with the typical bird arrangement of three toes pointing forwards and one back, but they share this feature with the large tribe of spinetails and swiftlets within the Apodinae. The remaining swifts, including our own have remarkably flexible feet, the four toes usually all pointing forwards but in opposed pairs when young (Chapter 7), and able to grip on a vertical wall by contracting between these pairs. These specialised feet have a strong grip, which is an irony seeing as the Latin name *Apus* derives from Aristotle and means 'no feet'.

Swifts range in size from the pygmy swiftlet, just 9 cm long and 5.4 g in mean weight to the purple needletail, 25 cm long and with a mean weight of 180 g. Both of these are found in the Philippines, the needletail in Sulawesi as well. Our swift is around 16 cm long weighing 36–50 g. The largest European swift, the Alpine, is around 22 cm long and weighs up to 90 g.

THE SWIFT SPECIES (SEE CHAPTER 15)

The common swift and its near relatives have been re-classified with the greater knowledge that we now have both on visible characters and using DNA. The common swift's distribution across Eurasia and North Africa, including subspecies *pekinensis*, remains as described in Chapter 15, but all the similar species further south have been separated as species, and there are now considered to be a total of eight species on the same evolutionary line (clade). One of these, the pallid swift, *Apus pallidus*, has long been separated from the common swift. It breeds in southern Europe as well as parts of North Africa, including the Canaries and Madeira, and the Middle East. The common and pallid swifts overlap considerably in the breeding range (Fig. 20, p.173 and Fig. 26) and, to an extent, in their winter range. None of the others in this clade migrates.

The plain swift, *Apus unicolor*, of Madeira and the Canaries, and the Cape Verde swift, *A. alexandri*, are now regarded as separate species. The African black swift, *A. barbatus*, has a patchy distribution but is known to be more widespread than recorded in Chapter 15. The Nyanza swift, *A. niansae* now includes *somalicus* as a race. Within the clade are also two swifts formerly seen as races of the pallid, Bradfield's swift, *A. bradfieldi*, of south-western Africa, and Forbes-Watson's swift, *A. berliozi*, unknown when my father wrote this book, first described as a species in 1966 from the horn of Africa and Yemen. There is only a little overlap in the breeding ranges of any of these in eastern Africa and in the Canaries and Madeira, but in the northern winter the common swift may be seen anywhere in Africa south of the Sahara (Fig. 18, p.166) and the pallid winters across tropical Africa north of the equator. All of these species are hard to distinguish in the field, and often hard to distinguish from other swifts, so there can be considerable difficulty in identifying swifts in Africa.

The Pacific white-rumped swift, described as having 'striking racial variation' in Chapter 15 has also been split, although some still

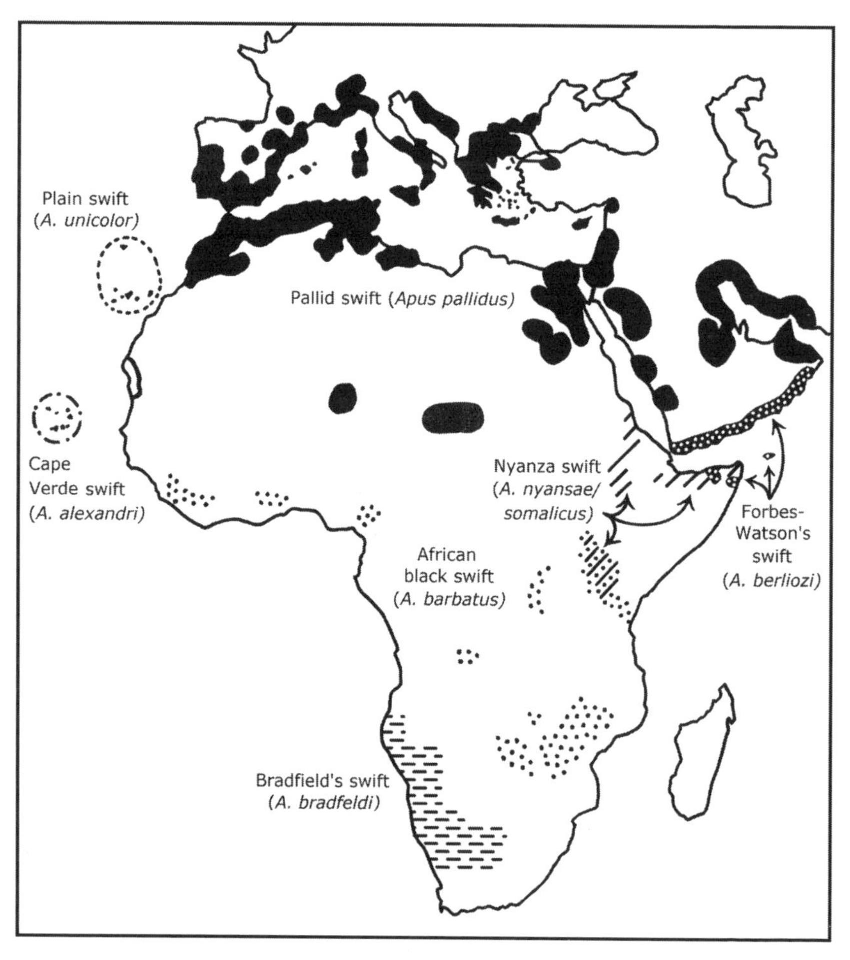

Fig 26: Breeding ranges of the seven closest relatives of the common swift (for range of common swift, including ssp. *pekinensis*, see Fig. 20, p.173)

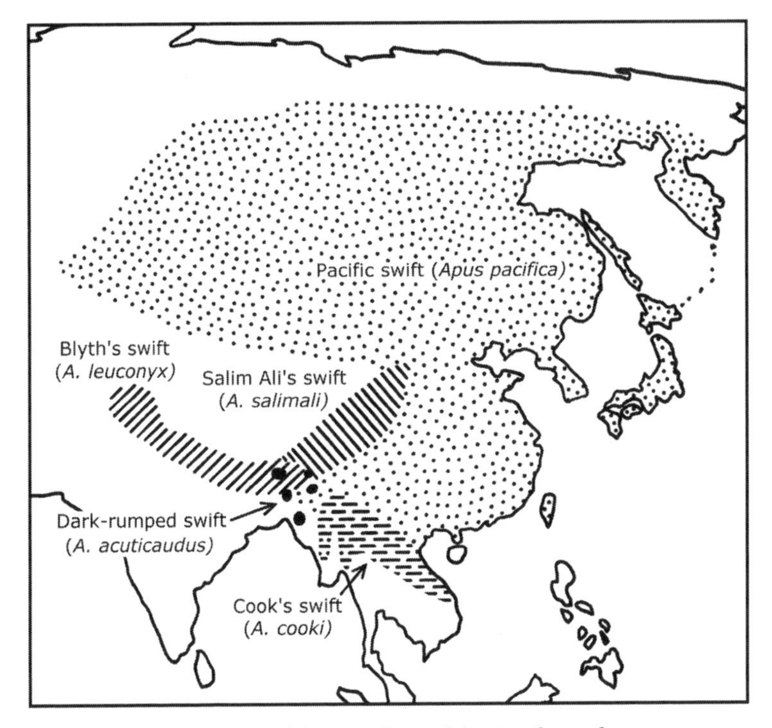

Fig 27: Breeding ranges of the members of the Pacific swift super-species

regard these as subspecies (Fig. 27). *A. leuconyx* (Blyth's swift) and *A. cooki* (Cook's swift) are seen as species, though regarded as races in 1956. The one referred to as ssp. *kanoi* from Taiwan in Chapter 15 is regarded as a subspecies, although birds from the Tibetan plateau and nearby also had this name. These are now regarded as a separate species, but the name *kanoi* was confusing and not valid. The species has been named *Apus salimali*, or Salim Ali's swift, a name that would have pleased my father greatly, as named after his friend, the great Indian ornithologist Salim Ali. These are regarded as forming a superspecies or species complex. The smaller, darker, dark-rumped swift, *Apus acuticauda* (or *acuticaudus*) is confirmed as a different species with a very restricted range and probably no more than a thousand individuals in total.

The European Alpine swift and the mottled swift of Africa, also mentioned in Chapter 15, have been removed from the genus *Apus* to be placed in a separate genus, *Tachymarptis*.

Swifts are largely tropical and only a few migrate northwards into temperate Eurasia, a rather surprising feature since distance would not appear to be a problem for birds that fly so much anyway, and those that do can take advantage of the long temperate summer days. But it seems that more swifts are spreading north. Five species of swift have now been recorded breeding in Europe. In addition to the common, pallid and Alpine swifts, the white-rumped swift, *Apus caffer*, was proved to be breeding for the first time in 1966, and the little swift, *Apus affinis*, in 2000, both in southern Spain. Both only breed in very small numbers, although the white-rumped swift has increased to around 100 pairs and extended to mid-Spain and Portugal. There are probably only about 30 breeding pairs of little swifts. The little swift is the only European swift to spend the winter in Europe. Both these colonisations follow range expansions within Africa. Bird distributions are constantly changing and we could well see more change in swift distributions in the next few decades, perhaps with increasing colonisations from Africa.

NESTING

Swifts, like many migrating birds, return to the same area, and usually the same nest site or very close to it, to breed, despite the enormous distances travelled between breeding seasons. The ringing of adult swifts in the tower has been limited, after the early finding that established pairs may desert if disturbed (Chapter 2), but Roy Overall and George Candelin have successfully ringed breeding adults by waiting until the young are almost ready to fledge; it is considered risky to ring single adults sheltering in a box and those that come in when breeding has finished. Between 1962 and 2010 they ringed 674 adults. 291 of these, 43%, have been re-trapped in the tower subsequently. Many are clearly faithful to the nesting site. Once they reach adulthood swifts can live for twenty years, though the average is around eight years. The oldest recorded in the tower was ringed as an adult by my father in June 1948 and re-found, recently dead, by Roy Overall in 1964, sixteen years later, meaning that it must have been at least eighteen years old, probably twenty. My father recorded the swift as being the longest-lived of all birds studied except the yellow-eyed penguin (Chapter 17), but several sea-birds, a few waders such as oystercatchers, parrots and birds of prey are now known to live longer; a Laysan albatross female has been recorded rearing a chick when at least sixty-six years old. The swift remains the longest-lived bird of its size.

Recoveries of fledglings tell quite a different story. Over the years, Roy ringed 4,064 young birds that fledged from the nest, of which only 52, 1.3%, have been re-trapped in the tower. It seems that, although the majority of these will have died within their first two years of life, i.e. before they bred for the first time, some must have nested elsewhere, and perhaps young birds do disperse to new breeding sites. Otherwise this is unsustainably low. What evidence we have suggests that most pairs raise their first young to fledging only when the adults are at least three years old, though they may prospect for nesting holes in their first summer after fledging and attempt to breed a year later. As in all birds, the most vulnerable time in their lives is the first year.

Swifts normally lay two or three eggs. The average clutch size over all the years recorded by Roy Overall was 2.1 eggs. Clutch size was a subject that preoccupied my father for many years during and after his part in the swift study, but he was always reluctant to manipulate or experiment with wild birds, preferring straight observation. In the late 1980s to early 1990s Thais Martins did manipulate the broods in the tower under the supervision of my father's successor, Prof. Chris

Perrins. She followed broods of one, two, three and four eggs. What she found was that, on average, more young fledged from larger broods. Clearly the number fledging is only part of reproductive success. Early observations showed that, even with only two young in the nest, the adult swifts can have difficulty finding enough food in poor weather, hence the ability of the nestlings to lose heat and go torpid (Chapters 7, 8). In Martins' study, the number of visits by the adults did increase with brood size, but not in proportion to the number of young. Those adults feeding large broods lost more weight themselves, some perhaps to critically low levels that may have affected their own survival. Some stopped bringing food in. Many swift nestlings weigh more than the adults before they leave the nest and then lose weight while their wing feathers complete their growth before fledging. In Martins's broods of one, the fledglings did end up heavier than those from larger broods, but the chick lost more weight before fledging.

The young may have fledged and the adults flown away on their migrations but this is not the crucial test of reproductive success. That has to be lifetime reproductive success, so far not possible to measure among swifts owing to the paucity of ringing recoveries of fledglings. There is a suggestion of an effect from a study by D. L. Thomson and others. They found that, although survival and growth of the nestlings was mainly correlated with the weather in June, poor weather in July led to lower survival of the adults through to the following year. Perhaps they could not build up enough reserves for migration. With a long-lived bird this kind of difference may be crucial. Early clutches appear to be at an advantage and, if a second clutch is laid after loss of the first for any reason, the second clutch is nearly always smaller than the first. It seems that, as in all birds, the mean number of eggs laid maximises the chances of successful lifetime reproductive success.

Despite the years of study we are still not much closer to knowing why swifts, at times, will remove eggs from their own nests. Poor weather and lack of food is undoubtedly one reason, and fights with intruding adults can stimulate ejection. All workers have found that, if they replace an egg that has once been removed by the adult swifts, the swifts will normally remove it again.

SCREAMING AND BANGING

The screaming call is one of the most obvious characteristics of the swift. Once a pair establishes itself in a nest hole they will sit at the entrance together and scream in duet at potential intruders (Chapter 3). The idea that there may be a difference between male and female screams has been suggested several times. An insight has come from Konrad Ansorge using a spectrogram. The number of recordings he made has been very small but he has shown clearly that both male and female screamed in sequence at around the high-pitched 6000 Hz point. The details differed between birds; any one bird could vary the frequency depending on the intensity of the stimulus and either sex could have the higher-pitched scream. Each scream ended with a rhythmic trill and here there was a consistent difference between male and female in the three pairs he recorded. The mean time between trill notes for the females was 26.9, 29.8 and 29.1 milliseconds, and for the males was 17.9, 18.1 and 16.7 milliseconds, a significantly faster trill that did not vary in frequency with intensity of screaming. Each bird was very consistent in its trills and there were significant differences between the individuals, but always the females had a much slower trill than the males. Since the sexes of swifts look alike this may be one way that the swifts themselves can tell the sexes apart.

A feature of swift behaviour that has given rise to some speculation is the 'banging' behaviour described in Chapter 3. My father's tentative explanation (Chapter 3) centred on birds prospecting for unoccupied nest holes. The banging by a passing bird stimulated the occupants of the hole to scream. Grzegorz Olos studied this behaviour in Poland and found something rather different. He showed that the banging happened when one or more birds had just settled in a nest hole or on the wall outside and another, the banger, was flying with them. It banged but did not settle, though the banging disrupted its flight. Almost all the banging was done within two seconds of the first bird(s) landing and the frequency of banging increased with the number of swifts flying around. There was an increase in banging behaviour through the breeding season too, with a marked peak in July when there were young in the nest. Frequently it appeared that when a pair approached the nest together, one would settle and the other bang. In these situations the settling of one bird and the banging of the second bird coincided.

One thing Olos found in his colonies was at least two pairs of kestrels catching swifts as they landed by their nests, the only situation where a kestrel would be able to catch a swift. He found a highly significant negative correlation between banging and predation, i.e. the more banging there was the fewer hunting attempts and fewer successful hunts there were by the kestrel. He interpreted the banging as, at least mainly, an anti-predator behaviour. We must be cautious at this interpretation, because the correlation could arise because the kestrels found hunting harder when more swifts were flying about. Banging remains a little understood behaviour and there may be more than one function in different circumstances.

DEATH AND ITS CAUSES

Generally, birds of prey are not thought to be a major cause of mortality, although there are records of various species taking swifts at least occasionally (Chapter 17). We have never seen kestrels hunting swifts at the tower, unlike in Olos' study. There was persistent predation at the tower in 2008 when Roy Overall and the museum staff recorded a sparrowhawk waiting by the nests on many occasions, and again in 2010 when a number of breeding adults were taken by a sparrowhawk. Sparrowhawks are specialist bird feeders but mainly hunt using surprise. Roy watched one in 2008 taking a swift just as the swift flew up to its nest, no mean feat considering how fast swifts fly, although they do slow down markedly when near their nest holes. This is likely to have been its usual tactic. We do not know whether there was any association with banging. Another bird of prey, the hobby, has increased considerably throughout southern England since the 1960s. Hobbies frequently catch swallows and martins and sometimes swifts. Two swift rings were found in debris at the base of a hobby's nest in the north-eastern part of Oxford in 2007 and two more in 2008. These were from young birds that had recently fledged from the tower colony. Despite this, it is unlikely that hobbies are major predators of swifts.

The common flat fly or louse-fly parasite on swifts, *Crataerina pallida*, forces itself onto the attention of anyone studying swifts both by its abundance and its size (Chapter 17). The question as to whether it does the swift any harm was addressed directly by Mark Walker and Ian Rotherham. They manipulated the numbers of flat flies in swift nests, and compared those nests with enhanced parasite loads (mean maximum number seen $10.17 + 5.1$), with those with reduced loads (mean maximum number seen $3.00 + 2.53$). Their sample size was small but the results suggestive. What they found was that the visitation rate of the parent birds was similar, but that those whose nests had high parasite loads spent longer at the nest in each visitation and longer caring for the nestlings. There was no detrimental effect of parasite load on any aspect of breeding success from clutch size to fledgling number or fledgling weight. This may not be the most important impact. There have been many reports of grounded or dead swifts, nearly always fledglings, in late summer with many of these parasites on them. It may be that they can fledge, but are weakened by a heavy parasite load.

A more detailed study on the Alpine swift flat fly, the closely-related *Crataerina melbae*, showed, from experimental manipulations of parasite load in the nests, that the rearing period was prolonged by a mean of three days in heavily infested nests. The rearing period is normally 50–70 days, mainly depending on weather, so this does not appear to be a large difference, but they found that those nests heavily infested one year produced, on average 26% fewer young to fledging the following year, with this effect greater if they manipulated the brood to give a heavy parasite load in the second year as well. Swifts are long-lived birds and overall lifetime reproduction may well be affected by parasite load. The flat flies overwinter in the nests as pupae and the birds are usually faithful to particular nest holes, so parasitism could be similar from year to year.

FLIGHT

Swifts are supreme on the wing and, after all, have been named after their mastery of speed in the air. In his introductory paragraphs to Chapter 10 my father quoted from the poets Lord de Tabley, Walter de la Mare and Laurence Whistler but 'except for these brief mentions, the mastery shown by the swift in the air has not received the admiration it deserves.' Ted Hughes could almost deliberately have risen to that challenge with his celebrated poem, *Swifts*, published as part of his *Season Songs* in 1975 and vividly evoking their flight. It begins:

> Fifteenth of May. Cherry blossom. The swifts
> Materialize at the tip of a long scream
> Of needle. 'Look! They're back! Look!' And they're gone
> On a steep
> Controlled scream of skid...

Later he mentions 'their whirling blades'; 'their too much power' and 'their arrow-thwack into the eaves', all observations familiar to anyone who has watched swifts.

On a more prosaic note, we have now some accurate measurements of what that speed is from Professor Anders Hedenström and his co-workers from Lund University in Sweden. During migration or on feeding flights, swifts will normally fly at around 10 m s^{-1} (22 mph) but during 'screaming parties' in particular, they can fly much faster than that. Using two high speed cameras, placed 1,400 mm apart on a beam, they recorded swifts flying at speeds of 11.9-31.1 m s^{-1} (26.6-69.6 mph) with an average of 20.9 m s^{-1} (46.8 mph) in screaming parties on a calm clear day at a local colony. At the same time these birds were actually rising upwards in the air at an average of 4.0 m s^{-1} (9 mph). These are only short bursts, but powered flight of these astonishing speeds has never been accurately recorded in any other bird. It is certainly possible that other swifts may be able to fly faster than our swift but this needs confirmation using accurate methods of measurement; speeds such as the 47 m s^{-1} (105 mph) mentioned for the white-throated needletail swifts are almost certainly exaggerations. Accurate measurements of ducks, other contenders for the fastest of birds in powered flight, have shown a relatively sedate 22-24 m s^{-1} (49-54 mph) maximum. A peregrine and other falcons in a stoop may still be the fastest birds in the air, and the speed record for any animal, although here again there have been exaggerations, accurate measurements suggesting a maximum speed of 51 m s^{-1} (114 mph) for a peregrine, 58 m s^{-1} (130 mph) for a trained gyr falcon, but that is using the full force of gravity and not as powered flight.

The Lund team measured wing-beat frequency as well, finding 7-8 beats per second during migratory or roosting flight rising to an average of 10.4 during screaming parties, with a range of 9.1–12.5. There was a weak relationship between speed of wing beat and flight speed. In further experiments, using a wind tunnel, they found that swifts varied their wingspan and tail span during a glide, with narrower wings and tail in higher wind speeds.

SLEEP

Swifts only roost in buildings, or rarely on trees, when breeding, prospecting for breeding holes or occasionally in the autumn before migrating. Are they then airborne for the entire period between breeding seasons, i.e. eight or nine months of each year, never touching solid ground? Sleep is thought to be essential for all higher vertebrates so this does seem extremely unlikely. We know that other species of swift such as the North American chimney swift do roost at night (Chapter 11). Night activity of swifts defied study for many years. They behaved so differently from other birds by flying high into the air at dusk rather than finding a roost like other birds. In the last decade there have been several insights using micro-electronics, particularly from Hedenström and his co-workers. They have managed to equip swifts with micro-dataloggers including an accelerometer and geolocators and showed that swifts do indeed spend up to 300 days on the wing without landing at all.

The occasional juvenile swift roosted in a tree or a building during bad weather soon after fledging in its first autumn, and occasionally a bird could have settled in the winter quarters during bad weather, but at least 99% of the time between breeding seasons was spent on the wing. They recorded very occasional bouts of flight inactivity of more than seventy minutes, the dataloggers registering that the bird was upright so not flying, but the 'inactivity' of short durations was most probably gliding. Even most fledgling swifts, from their first time out from the nest, did not land again until they returned to their breeding area the following year. No wonder the nestlings are hesitant before their first flight.

The Swedish workers also confirmed that the visual accounts and early radar study of swifts flying up at twilight, the so-called 'Vespers flight' during the breeding season, were accurate, that the birds stayed up all night, and that this happened in their winter quarters as well as their summer quarters if they were not on a nest. Indeed, in their winter quarters the swifts flew up to around 2500 m at dusk, usually 6–7 p.m. and then again at dawn, around 7–8 a.m. The reason for this is not known as it did not appear to be for foraging in the dark. Perhaps it helps the birds orientate or helps them stay away from predators. Only when breeding would they settle and then spend the night in the nest.

Common swifts, it turns out, are not the only birds that can stay active all night. Dr Niels Rattenborg at the Max Planck Institute in Seewiesen, Germany, has identified that species from six different orders of birds undergo long-distance migrations that require non-stop flights for more than twenty-four hours. These include several waders (shorebirds), with the bar-tailed godwit holding the geographical distance record by flying from Alaska to New Zealand in an average of 8.1 days, a speed that means they cannot stop overnight. Others include waterfowl, birds of prey, albatrosses, frigate-birds, a few passerines and possibly others. Only swifts spend months in the air, but it turns out that Alpine swifts do this outside the breeding season as well as common swifts, spending up to 200 days airborne at a stretch. It is likely that some other swift species do this too. The question remains as to how much of this time is actually spent sleeping.

Rattenborg and his colleagues studied the great frigate-birds of the Pacific and Indian oceans to shed some light on this. These large lightly-built seabirds soar a great deal and rarely flap. Young birds have been shown to spend more than two months continuously on the wing cruising around the Indian Ocean. The researchers managed to implant sensors surgically into the brain and attach a datalogger on the heads of some breeding frigate-birds so that they could determine not only whether they were asleep or awake but also the type of sleep they were having. This showed that the frigates, like dolphins, appear to be able to sleep asymmetrically, with half of their brain in deeper sleep than the other half, and with one eye closed. This allows dolphins to keep swimming and frigates to keep gliding. Curiously the frigates, for brief periods, did actually sleep with both hemispheres at once while circling and even had rapid-eye-movement sleep for up to five seconds at a time. Despite this the frigates slept very little when airborne, averaging only 0.7 hours per 24 hours. On land they slept for an average of 12.8 hours per 24. The foraging trips that they studied lasted an average of 5.8 days, with the longest being ten days, so we do not know whether the patterns may be different for longer spells. Asymmetrical sleep has also been shown among ducks when roosting on land, keeping them vigilant.

We do not know how relevant the frigate-bird study is for swifts or other birds. It does seem to indicate that swifts, like the frigates, may not flap when asleep but only glide. Studies on swifts have shown that, at night, they alternate between flapping and gliding with each episode lasting for up to six seconds. Swifts will orientate themselves into a wind at night but appear not to adapt their flight speed with the strength of the wind. Their night flying speed was similar to that during the day at about 9.6 m s^{-1}. This means that they will be displaced much further in a strong wind than a light one. A curious finding was that the direction they faced was not always directly into the wind but oscillated during the night either side of the wind direction with an apparent periodicity, though the reason for this is obscure. In a very light wind they flew in random directions with no clear orientation. Swifts fly at much higher altitude at night, one experiment involving fourteen swifts fitted with altimeters showing an average maximum altitude of 2770 m by night, and 1700 m by day. The highest flying went up to over 3500 m.

Study of sleep patterns in a flying bird is clearly difficult, but with electronic devices becoming smaller and more accurate all the time it is likely that we will gain more insight in the coming years.

MIGRATION

Common swifts are found across much of Eurasia in the summer, but all spend the northern winter in Africa south of the Sahara.

Here they mix with other species of swift and it can be very hard to tell the species apart as they form mixed flocks. Most observations of the common swift come from the semi-arid interior of southern Africa, especially Botswana and neighbouring countries. This may be partly because there are more birdwatchers there and because birds are easier to see in open savanna than in woodland or forests. The countries with the most recoveries of British swifts are the two Congo countries and Malawi although the number is still small.

The British Trust for Ornithology (BTO) has placed some geolocators on British swifts, and Susanne Åkesson, Anders Hedenström and their Lund team have done this successfully with many more. They tracked swifts to their wintering grounds in Africa. The Lund workers equipped 157 adult swifts with these geolocators from six different colonies in Sweden, one in the north, two near the centre and three in the south of Sweden. They could eventually use 72 of these swifts to determine where they went. There were no discernible differences between the colonies. The swifts set off from Sweden in a south-westerly direction to reach the Mediterranean via Iberia. Many stopped in Iberia for two weeks or so before heading south to the Sahara. They crossed the western side of the Sahara in the autumn on a broad front, roughly from coastal Mauretania to the Niger/Chad border. They varied a lot in the length of time it took them, the quickest being four days, the longest 72.8 days, an average distance covered being well over 300 km per day. It seems that they would wait for a tail wind and make detours during this time. They wintered in the northern part of the swifts' winter range; across equatorial Africa between the latitudes of 8° N and 6° S. The British swifts so far tracked appeared to winter from the Congo countries to the eastern seaboard of Africa in Mozambique, so in the southern part of the range of the Swedish birds and similar to the ringing recoveries of British swifts. They move about across this wintering range during the winter.

In spring all travelled along one of three quite narrow routes, unlike in the autumn. Most, including all those that wintered in west Africa, gathered for a stopover of around ten days, presumably feeding up for the journey, in Liberia or nearby countries. They then flew north to cross the Sahara more quickly than in the autumn travelling along a western coastal route. Those that wintered in the Congo flew across the central Sahara. One bird went a long way east to the edge of Arabia. It seems that the swifts would usually wait for a tail wind and then cross the Sahara quickly, the quickest managing the 3700 km or so in two days with a favourable tail wind. The average length of time was 5.5 days; the longest recorded 13.7 days. Those crossing the central route were the quickest, perhaps because of lack of feeding opportunity on the way. They set off between 26 April and 28 May (average around 10 May).

In 2014 similar geolocators were placed on 31 'Beijing swifts', subspecies *pekinensis*, in Beijing itself. This is slightly paler than the European common swift. Thirteen of these returned the following year and demonstrated that they started their migration by flying westwards and even slightly north into Mongolia, taking a route north of the Tianshan mountains, before turning south-west through Iran, stopping for a few days by the southern part of the Caspian Sea, then crossing central Arabia and tropical Africa to winter in Namibia and western South Africa. This is the southern part of the known winter range, and likely to be their normal winter range as different populations or races of species often winter, as well as breed, in different areas. They returned along a similar route. This migration is a distance of around 13,000 km (8000 miles) each way, though of course the swifts will fly much further than that in total. The distances a swift may fly in a lifetime are staggering. If one lives for twenty years it must fly around 2.6 million km (1.6 million miles), the equivalent of nearly seven times the distance to the moon.

Earlier work on the homing ability of swifts showed that when twelve swifts were caught at their colony in Sweden, transported and then released 405 km south, most, and probably all, returned by the following morning. Some were transported north another year and these took several days, presumably over unfamiliar territory, but there was bad weather then too that probably disrupted them.

Work using geolocators, and, most recently, miniature GPS-loggers which considerably improve the accuracy, is ongoing and the technology improving all the time, so we are likely to learn much more about the routes and timings of migrations over the next few years.

NUMBERS IN THE TOWER

Like any bird, swifts fluctuate in number from year to year, and there are good and bad seasons for raising young. Swift numbers are buffered better than many birds by the longevity of the adults. We can get some ideas about numbers from the long-term study in the tower (Fig. 28). For the graph of 'occupied nests' I have used the number of nests in which at least one egg was laid, for which there are clear results since 1991. Every year some adults roost in boxes near the start of the season and some start to build nests but abandon these before eggs are laid, so these are not included. A number more desert after one or more eggs are laid, usually in a period of bad weather; these are included. Roy Overall and George Candelin never disturb the adults at the start of the season and it is sometimes impossible to tell whether a sitting adult has an egg. This will nearly always become clear later, although the occasional egg may have been omitted from the results if it was ejected from the nest. Roy and George have ringed almost or quite all the nestlings in the tower at least four weeks after hatching and within two weeks of fledging, as by then they are sufficiently grown for a ring to stay on their legs. These provide a clear demonstration of numbers of young raised each year. The great majority of these fledge successfully. Since 2000 we have more definite numbers of birds fledged. These are slightly lower than the

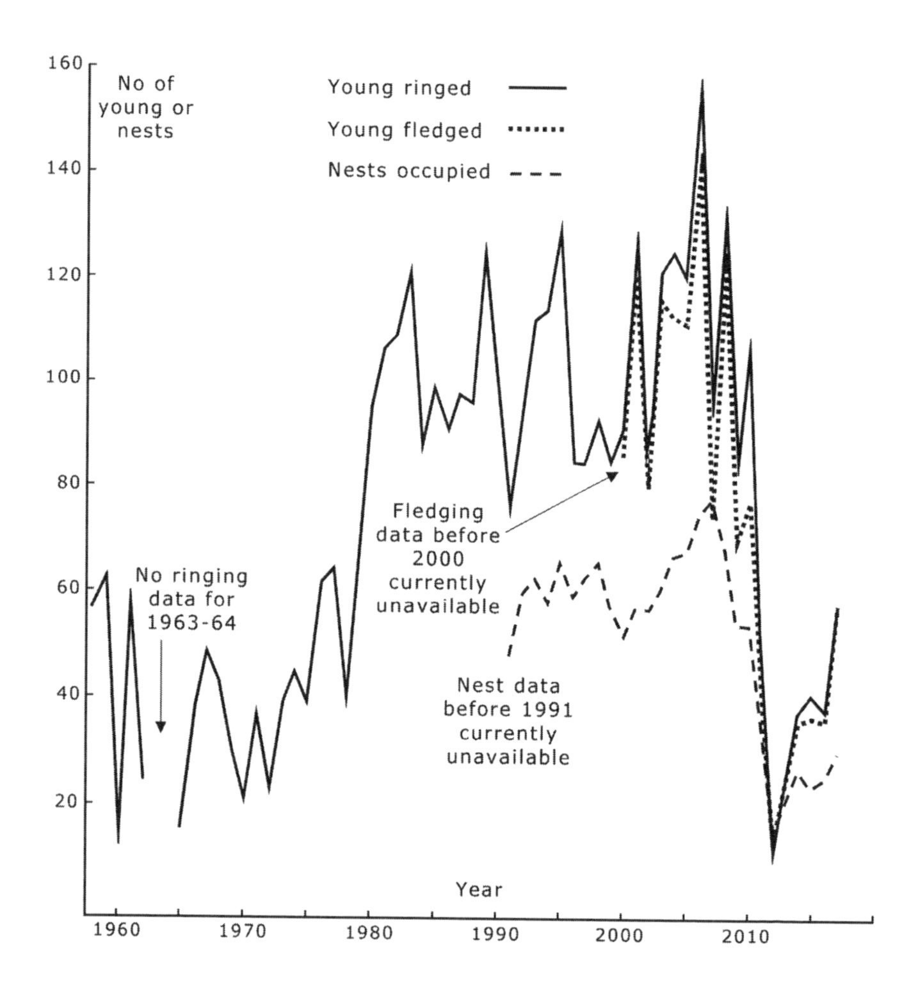

Fig 28: Numbers of young swifts ringed in the tower, numbers fledged and numbers of nests in which at least one egg was laid

number ringed in most years because of a few late deaths of young birds normally from starvation, especially in bad weather or, in 2010 and possibly other years, predation of the adults by a sparrowhawk.

When the study in the museum tower was set up, almost all of the 40 available boxes were occupied. Once the number of nest boxes had been increased to 147 for the 1966 season more boxes appeared to be occupied but no more young were produced in total. Indeed the number went down, with as few as 21 young in 1970. It is not obvious why there was the rapid increase between 1978 and 1981 but this was sustained, with fluctuations, at around an average of 100 young birds in total until 2010.

The largest number of young fledged in any one year in the tower has been 141 (156 ringed) in 2006, an exceptionally warm dry summer with the warmest July on record. This happened even though some young died in the heat. Indeed, the eight-year stretch between 2001 and 2008 was particularly successful for the swifts, mostly warm summers and with more than 120 young raised each year except in 2002 and 2007, both of which had poorer summer weather. Three other years with more than 120 young ringed have been 1983, 1989 and 1995, all years with warm sunny summers.

Starting in 2008, but mainly after 2010, there was a precipitous decline in numbers. From 2011 to 2016 the average number of young raised was 33.7 (range 13–48). This decline probably resulted from a combination of factors. The various disturbances to the tower in 2010-11 (see *The Tower* above) may well have led to some not returning the following year. This coincided with 2010, 2011 and 2012 all having dismal summers, with below average temperatures throughout June and July in all three years and 2012 being the wettest June on record.[2] 2012 produced the fewest young swifts on

[2] The Olympic Games in London that ran from 27 July until 12 August 2012 were extraordinarily fortunate, coinciding with almost the only fine weeks of summer 2012.

record; only 13 fledged young in total. It is likely to take time for the numbers to recover. Happily, 2017 had a very warm June and early July and it has been the best year since 2010, with 58 fledged from 25 nests (five further nests failed after eggs were laid), with many pairs fledging three young. This is still a lower number than any between 1979 and 2010, but the increase in numbers fledged is encouraging.

Only just over half of all available nest boxes have been occupied even in the best of seasons, but the large majority of those under the cowls have been occupied when the birds are present in good numbers. It is the boxes under the eaves that have never been popular. The topmost boxes in the tower are occupied first by returning swifts and there has been a slight, but significant, preference of the swifts for the easterly facing roof of the tower (mean 15.2 nests occupied per year; 14.1 on west roof, 13.1 on north, 13.0 on south). Perhaps the little extra warmth in the morning is appreciated, or the less direct radiation at the hottest times of day. The number of boxes in which at least one egg was laid does not correlate strongly with the numbers of young. The largest number of occupied boxes recorded for any year was in 2007, but this was a year with rainfall far above average with unprecedented flooding in late July in Oxford. There was a marked drop in the total number of young fledged. Clearly many swifts can start to build nests, but the weather has the largest influence on how many fledge.

One other striking feature of the swifts at the tower has been the first appearance of the swifts in the spring. Between 2007 and 2013 first sightings were all between 24 and 29 April, mean 26 April; from 2014-2017 first sightings have all been 4 or 5 May an average of nine days later. Only four years are involved so far and this could, of course, be chance, but it is consistent and comes just after a distinct decline in numbers.

CONSERVATION OF SWIFTS

When my father wrote *Swifts in a Tower*, there was no particular concern about swift numbers. Breeding as they do in roofs, of which there would be no shortage, and feeding on the abundant small invertebrates in the atmosphere, they would have seemed secure. But all workers on swifts agree that they have declined everywhere in numbers in the last 20-30 years. The BTO, in their 2007-11 Atlas estimated that they had declined by 38% in Britain between 1995 and 2010 and by 46% in Ireland between 1998 and 2010. The most recent report from the RSPB and other organisations suggests a 51% decline between 1995 and 2015 in the UK. There have been other figures quoted, mostly near these figures, depending on which location and exactly which years are being considered. In 2009 estimates ranged from 87,000 to 160,000 breeding pairs in Great Britain. In China the decline is estimated to be around 60% in Beijing and its surrounds since the mid 1980s. Swifts have an honoured place in Beijing society, having nested in the eaves of the palaces and gatehouses of the city for many centuries, so the decline is much lamented.

In the tower there is no shortage of nest sites and we can put the fluctuations in numbers to summer weather and, perhaps, disturbance. The longer term decline more generally across the whole of the swift's range must have come from some more general problems, and these have been much debated. Shortage of nest sites has been proposed as a major reason in some places. Swifts nest in holes in roofs, in thatch and under the eaves. The great majority of houses today are too well built to have holes in these places, and holes have mostly been blocked up in older houses. This is the reason behind the production of purpose-built swift boxes and swift bricks. Designs for these have proliferated in recent years and are now being produced by several companies, including the RSPB. They are increasingly being placed in buildings, including churches and factories as well as private houses. We have yet to see whether this will affect the overall numbers of swifts breeding. Because adult swifts are faithful to their nest sites, it is likely to be young birds that are attracted to a new site. The best way to encourage birds to investigate is likely to be playing recordings of the calls, suggesting that birds are already there, although anyone interested should obtain specific advice on welfare from the RSPB or websites such as @SaveourSwifts. This has been shown to be most effective when the non-breeders arrive, a bit later than established adults in late May and June. Swifts are using some of these artificial sites and they could well be important where potential nest sites are in short supply.

Many birds have declined in numbers, those associated with farmland being particularly affected in Britain and in all countries with intensive industrialised farming. The monocultures of industrial farming have few insects and a comparison of pasture land and cereal fields in Oxfordshire suggested that there were between two and seven times as many flying insects over pasture as there were over cereal crops. All aerial insectivores will be affected by this. It is possible that rape crops are better than cereals, as in recent years, study in the tower has shown that one important food for the young has been Brassica flea-beetles that can be abundant in rape fields (despite efforts by farmers to limit their numbers). There are many rape fields around Oxford every year.

In the USA the decline of aerial insectivores had a strong geographical gradient and an association with migration distance. The further the birds went to the north-east, the most industrialised part of the country, the greater the decline, though the trend extended further north-east than the most industrialised part. In addition the further the birds migrated the greater the declines, with those wintering in South America having 1.1% decline per year compared with those wintering in Central America averaging 0.25% per year. Both of these trends can be linked to the availability of flying insects in the breeding areas. Shortage of food will make the birds more vulnerable both in their breeding sites and on their long migrations.

Karl Evans and co-workers in Sheffield University used a quite different technique to find out where the birds had been feeding in winter. They analysed body feathers from breeding adult wood warblers, house martins and swifts for different isotopes of carbon. In tropical Africa, woodlands are dominated by plants with the usual three-Carbon (C3) metabolic cycle but grasslands are dominated mainly by grasses using four-Carbon (C4) fixation of the carbon. These two types give rise to recognisably different levels of the ^{13}C isotope of carbon in the plants and these get passed up through the food chain. These three bird species moult their body feathers in their winter quarters incorporating these isotopic differences into the new body feathers. The workers found that all the wood warblers had almost entirely C3 carbon so relied on woodland, house martins had a mixture of levels but very little difference between individual birds, suggesting that they were in a restricted area perhaps of savanna woodland. Swifts, on average, had similar levels of the isotope to the house martins but there was a great range of variation between individual birds. This suggests that, overall, swifts were utilising a wider area and more different vegetation types than either of the other species. This should make them less vulnerable than the more specialised wood warblers or house martins. All three have declined but, in Britain, the wood warbler has declined the most, an estimated 63% between 1995 and 2009. House martins have had mixed fortunes: they actually doubled in numbers in Scotland between 1995 and 2010 but declined in England by 14%, and by 40% in France between 2000-2010. Evans and his colleagues estimated that swift populations in Britain declined by 31% between 1995 and 2009, although the BTO estimate was 38% from 1995-2010 (see above). These numbers have not been reflected directly in the tower population at all.

Chimney swifts in North America have seen large declines, like our swift, and, for them there is no shortage of nesting sites, one study showing that fewer than 25% of suitable chimneys were actually occupied by the swifts. The conclusion here was that agricultural intensification and a shortage of insect food was the cause.

It will be interesting to see if the erection of a swift tower in Oxford or elsewhere will make a difference to breeding numbers. We should not expect the causes of the decline in swifts to be the same everywhere. Predation seems unlikely as a cause of decline for swifts, but availability of nest sites, food in the breeding area or anywhere along the migratory route or wintering quarters could contribute to the declines. Throughout its breeding range the decline of swifts is felt keenly by many people and we must do what we can to reverse it.

Having said all of that, the common swift remains the most numerous swift species in the world, with an estimated 40-200 million birds. The chimney swift, at around 15 million, is the most numerous in the Americas, and second only to the common swift overall.

CONCLUSION

The swift was known as the martlet in Medieval England and, because its name *Apus*, stemming from Greek times, means 'no feet' it has come into heraldry to symbolise the fourth son – one deprived of inheritance who has nowhere to put his feet. In our university city of Oxford there is a further, associated, meaning that has particular resonance. The heraldic martlet has been used as a symbol of the unceasing quest for learning, knowledge and adventure. This extra meaning has been taken up by colleges in both Oxford and Cambridge Universities. From its early foundations three Oxford colleges have incorporated martlets into their coats of arms: University College, founded in the 13th century, the 18th century Worcester College and the 20th century St. Peter's College. It makes it all the more appropriate that the pioneering study of swifts took place in Oxford.

'The swift is one of the most remarkable, yet one of the least known, of all British birds,' said my father to begin this book. Many people have become fascinated by swifts since then; people have accessed colonies of swifts; watched them; studied their habits and recorded them, some using technology that was inconceivable in 1956. The result is that a great amount has been discovered about swifts, at least partly inspired by his study. I think we can say now that the swift is actually one of the better known of British birds. It is certainly one of the most remarkable of our birds, and one of the most alien of our common town birds to our own lives. To quote Hughes' poem, 'the fine wire they have reduced life to' has given us an unending source of admiration and wonder.

ACKNOWLEDGEMENTS

Roy Overall and George Candelin unhesitatingly provided results, comments and their enormous knowledge of swifts and it is a great pleasure to dedicate this chapter to them. I am most grateful to Chris Perrins, Anders Hedenström, Colin Wilkinson, Helen Moorhouse, Sandra Bouwhuis and Andy Gosler for their comments, contributions and help in writing this chapter. This publication would not have been possible without the help of Charlotte Kinnear and staff at the RSPB and Lucy Duckworth and those at Unicorn Press. Particular thanks to Colin Wilkinson for providing the lovely cover picture and new and updated illustrations, to Steve Blain and to Manuel Hinge for his several photographs, all taken in the tower itself.

Andrew Lack

REFERENCES

Note: To save the general reader, references are not given in the text of the book. Those who wish to check the source of any observation will find it under the appropriate chapter and page number in the following list. Where more than one reference occurs on the same page of the book, further guide as to which is meant is provided by the title of the work or by a note of the point concerned added in brackets after the reference. Where one paper summarises many others, I have omitted the others unless of special interest. I have occasionally omitted reference to my own papers and to the comprehensive study by E. Weitnauer first cited below under pp.20–21.

CHAPTER 1

P.13 WHITE, GILBERT 1795. *The Natural History of Selborne* (especially letter 21 to the Hon. Daines Barrington, dated 28 Sept. 1774.

P.15 LACK, D. 1956. A review of the genera and nesting habits of swifts. *Auk* 73: 1–32 (numbers of foreign swifts).

P.16 TINBERGEN, N. 1953. *The Herring Gull's World* pp. 231–232.

P.16 For good examples of the 'bird-book swift' see H. F. Witherby's *The Handbook of British Birds* 1938, A. Thorburn's *British Birds* 1925, T. A. Coward's *The Birds of the British Isles and their Eggs* 1920, and many others.

CHAPTER 2

P.18 VERNON, H. M. & K. D. 1909. *A History of the Oxford Museum.*

P.18 HUXLEY, L. 1908. *Life and Letters of Thomas Henry Huxley.*

P.18 PUSEY, E. B. 1878. *Un-science, not Science, adverse to Faith.*

P.20 WEITNAUER, E. 1947. Am Neste des Mauerseglers, *Apus apus apus* (L.). *Orn. Beob.* 44: 133-182.

CHAPTER 3

PP.27–35 LACK, D. & E. 1952. The breeding behaviour of the swift. *Brit. Birds* 45: 186–215 (for general information on this chapter).

P.27 WEITNAUER, E. 1947. Am Neste des Mauerseglers, *Apus apus apus* (L.). *Orn. Beob.* 44: 133–182 (for general information on this chapter).

P.29 MOREAU, R. E. 1941. 'Duetting' in birds. *Ibis* 1941: 176–177.

P.29 LAYARD, E. L. 1875-84. *The Birds of South Africa* (new ed.) p.385 (bokmakierie).

P.29 MATHEWS, G. M. 1921–2. *The Birds of Australia* 9: 245–246 (coach-whip bird).

P.29 CHISHOLM, A. H. 1948. *Bird Wonders of Australia* (3rd ed.) p.200 (coachwhip bird).

PP.36–7 *Swifts competing for nests with other species:*

STENHOUSE, J. H. 1931. Swift versus starling and sparrow. *Scot. Nat.* 1931: 73–78 (reviews many other published instances).

HAMILTON, D. 1928. Further observations on the swift. *Scot. Nat.* 1928: 175–179.

JOURDAIN, F. C. R. 1901. On the breeding habits of the swift in Derbyshire. *Zool.* 4: 5: 286–289.

KIRKMAN, F. B. et al. 1910. *The British Bird Book* 2: 353–361.

SUMMERS-SMITH, D. & M. 1951. Starlings attacking swifts at nest-site. *Brit. Birds* 44: 89–90 (likewise *Brit. Birds* 44: 407, notes by H. E. Pounds, C. M. Morrison).

DAMM, W. 1934. Kämpfende Mauersegler. *Beitr. Fortpfl. biol. Vög.* 10: 191 (reported as two fighting swifts, but description suggests starling versus swift).

PRICE, F. W. 1888. Swifts laying in martins' nests. *Zool* 3:12: 68 (likewise in *Zool.* 3:11: 348, 391, 428).

SYMES, J. H. 1925. Swift brooding young house-sparrows. *Brit. Birds* 19: 177.

BERTRAM, K. 1906. Beobachtungen über *Apus apus* (L.). *Ornith. Monatsschrift* 31: 238–253, 257-260.

P.37 ANSLEY, H. 1954. Do birds hear their songs as we do? *Proc. Linn. Soc. New York* 63–65: 39–40.

CHAPTER 4

PP.38–45 LACK, D. & E. 1952. The breeding behaviour of the swift. *Brit. Birds* 45: 186–215 (for general information on this chapter).

P.38 LACK, D. 1940. Pair-formation in birds. *Condor* 42: 269–286 (a general review).

P.40 LACK, D. 1943. *The Life of the Robin* Ch. 5.

P.41 WEITNAUER, E. 1947. Am Neste des Mauerseglers, *Apus apus apus* (L.). *Orn. Beob.* 44:133–182 (notes on pair-formation, also copulation on nest).

P.41 CLOGG, S. 1883. Swift returning to former nesting place. *Zool.* 3: 7: 257.

P.43 TINBERGEN, N. 1952. 'Derived' activities; their causation, biological significance, origin and emancipation during evolution. *Quart. Rev. Biol* 27: 1–32.

PP.44–6 *Copulation*:

WHITE, G. 1795. *The Natural History of Selborne*, letter 21 to Daines Barrington.

TOMES, R. F. 1853. Note on the copulation of swifts. *Zool* 11: 3943–4.

Further records of aerial coition in *Beitr. Fortpfl. biol Vög.* 1929, 5: 112; 1931, 7: 229; 1932, 8: 73–74, 219; 1933, 9: 55-56, 221–222; 1934, 10: 193, 214–216; 1943, 19: 166, 171; also Slijper, H. J. 1948 Over de Gierzwaluw, *Apus a. apus* L. *Ardea* 36: 4251; Nicholson, E. M. 1951 *Birds and Men* p.219; Boreham, H. J. (*in litt.*).

(CHAPMAN, A. 1928. *Retrospect* p.312 also claims aerial copulation in lapwing and lesser kestrel, but no one else has done so.)

BERTRAM, K. 1906. Beobachtungen über *Apus apus* (L.). *Ornith. Monatsschrift* 31: 238–253, 257–260 (copulation on nest).

P.47–8 *Copulation in other species of* Apus:

SMITH, S. 1950. Some notes on the Alpine swift. *Brit. Birds* 43: 122–123.

ARN, H. 1945. Zur Biologie des Alpenseglers *Micropus melba melba* (L.). *Schweiz Arch. Ornith.* 2: 137–181 (Alpine swift).

MOREAU, R. E. 1942. The breeding biology of *Micropus caffer streubelii* Hartlaub, the white-rumped swift. *Ibis* 1942: 27–49.

HUXLEY, J. S. *in litt.* (aerial copulation of house swift).

MOREAU, R. E. 1941. A contribution to the breeding biology of the palm-swift, *Cypselus parvus. J. East Africa & Uganda Nat. Hist. Soc.* 15: 154–170.

SPENNEMANN, A. 1928. *Collocalia esculenta linchi* (Horsf.). *Beitr. Fortpfl. biol. Vög*, 4: 53–58, 98–103 (cave swiftlet).

BENT, A. C. 1940. Life histories of North American cuckoos, goatsuckers, hummingbirds, and their allies. *U.S. Nat. Mus. Bull. 176:* 254–319 (chimney, black and white-throated swifts).

BRADBURY, W. C. 1918. Notes on the nesting habits of the white-throated swift in Colorado. *Condor* 20: 103–110.

CHAPTER 5

PP.49–68 LACK, D. 1956. A review of the genera and nesting habits of swifts. *Auk* 73: 1–32 (a review with numerous references to all species).

The following references are additional to this review:

P.49 FORREST, H. E. 1907. *The Fauna of North Wales* p.193 (nesting in cliffs).

P.52 JØRGENSEN, J. 1950. Mursejlere (*Apus apus*) 'fanger' staniolstrimler. *Dansk Orn. For. Tidsk.* 44: 45 (taking tinfoil; also in letter from Italy during war by P. A. Clancey).

P.57 RAY, J. 1678. *The Ornithology of Francis Willughby* p.215.

P.56–7 WANG, C. C. 1921. The composition of Chinese edible birds' nests and the nature of their proteins. *J. Biol. Chem.* 49: 429–439.

P.57 BANKS, E. 1949. *A Naturalist in Sarawak* pp.5–12.

P.58 GRIFFIN, D. R. 1953. Acoustic orientation in the oil bird, *Steatornis. Proc. Nat. Acad. Sci.* 39: 884–893.

Appendix:

P.69 LACK, D. 1956. A review of the genera and nesting habits of swifts. *Auk* 72: 1–32 (a revised classification of the genera, differing in various ways from that of Peters, T. L. 1940 *Check-List of Birds of the World* vol. 4).

CHAPTER 6

PP.71–78 LACK, D. & E. 1952. The breeding behaviour of the swift. *Brit. Birds* 45: 186-215 (for general behaviour in incubation; data on the influence of weather on laying will be published later in the *Ibis*).

P.74 TINBERGEN, N. 1951. *The Study of Instinct* pp.113–119 (displacement activities).

P.76 MATTHEWS, G. V. T. 1954. Some aspects of incubation in the Manx shearwater *Procellaria puffinus,* with particular reference to chilling resistance in the embryo. *Ibis* 96: 432–440.

P.76 MOREAU, R. E. 1942. The breeding biology of *Micropus caffer streubelii* Hartlaub, the white-rumped swift. *Ibis* 1942: 27–49 (removal of egg).

Incubation periods of African species:

P.77 MOREAU, R. E. 1942. The breeding biology of *Micropus caffer streubelii.* Hartlaub, the white-rumped swift. *Ibis* 1942: 27–49; 1942 *Colletoptera affinis* at the nest. *Ostrich* 13: 137–147; 1941 A contribution to the breeding biology of the palm-swift, *Cypselus parvus. J. East Africa & Uganda Nat. Hist. Soc.* 15: 154–170.

CHAPTER 7

P.80 HEINROTH, O. & M. 1931. *Die Vögel Mitteleuropas* vol. 4. Nachtrag p.88 (blackcock can fly at 10 days).

P.80–1 PORTMANN, A. 1950. Le développement postembryonnaire. *Traité de Zoologie* ed. P. P. Grasse vol. 15 *Oiseaux* p.523 (comparison of young alpine swift and young quail).

P.83 STRESEMANN, E. 1927–34. *Aves* p.837 (down in crested and palm swifts).

P.85 RICHDALE, L. E. 1943. The white-faced storm petrel. *Trans. Roy. Soc. New Zealand* 73: 217–232 (growth of young).

P.85 RICHDALE, L. E. 1945. The nestling of the sooty shearwater. *Condor* 47: 45–62.

P.85 LACK, D. & SILVA, E. T. 1949. The weight of nestling robins. *Ibis* 91: 64–78.

P.85 HUGUES, A. 1907. Le jeune chez le martinet. *Bull. Soc. Zool. Paris* 32: 106–108.

P.87–8 Feeding behaviour of three African species, the palm swift, white-rumped swift and house swift: Moreau, R. E. 1941 *J. East Africa & Uganda Nat. Hist. Soc.* 15: 154–170; 1942 *Ibis* 1942: 27–49; 1942 *Ostrich* 13: 137–147.

P.88 DEXTER, R. W. 1952. Extra-parental co-operation in the nesting of chimney swifts. *Wilson Bull.* 64: 133–139.

P.88 SKUTCH, A. F. 1935. Helpers at the nest. *Auk* 52: 257–273.

CHAPTER 8

P.89 LACK, D. & SILVA, E. T. 1949. The weight of nestling robins. *Ibis* 91: 64–78.

P.91 KOSKIMIES, J. 1948. On temperature regulation and metabolism in the swift, *Micropus a. apus* L., during fasting. Experientia 4: 274–282.

P.92 PEARSON, O. P. 1950. The metabolism of hummingbirds. *Condor* 52: 145–152.

P.94 GROHMANN, J. 1939. Modifikation oder Funktionsreifung? *Zeits. Tierpsychol.* 2:132–144.

CHAPTER 9

PP.97–107 LACK, D. & OWEN, D. F. 1955. The food of the swift. *J. Anim. Ecol.* 24: 120–136 (gives references to all the other observations described in this chapter except those specified below).

P.104 MATOUSEK, B. 1951. Contribution to the biology of the bee-eater (*Merops apiaster*) in Slovakia. *Sylvia* 13:125.

P.105–6 HAARTMAN, L. V. 1949. Neue Studien über den Tagesrhythmus des Mauerseglers, *Apus apus* (L.). *Orn. Fentt.* 26:16–24 (times of rising and roosting).

P.105 SCHEER, G. 1949. Beobachtungen fiber den morgendlichen Flugbeginn des Mauerseglers, *Micropus apus* (L.). *Vogelwarte* 2 (15): 104-109 (times of rising and roosting),

P.106 ANON. 1860. *The Reason Why Natural History* by the author of *The Biblical Reason Why* (etc.) pp.235–236.

CHAPTER 10

P.108 DE TABLEY, LORD, i.e. WARREN, J. B. L. The Invitation, quoted from *The Collected Poems*, 1903.

P.108 DE LA MARE, W. 1943. Swifts, from *The Burning Glass.*

P.108 WHISTLER, LAURENCE 1940. Flight, from In Time of Suspense.

PP.109–12 General accounts of flight: *Encyclopedia Britannica,* Jack, A. 1953 *Feathered Wings*, Storer, J. J. 1948 The Flight of Birds (*Cranbrook Inst. Science Bull.* 28), Barlee, J. 1947 *Birds on the Wing*, Brown, R. H. J. 1951. Flapping flight. *Ibis* 93: 333–359

PP.109–12 LORENZ, K. 1933. Beobachtetes über das Fliegen der Vögel und über der Flügel- und Steuerform zur Art des Fluges. *J. f. Ornith.* 81: 107–236 (good comparative account of flight, including that of swift).

PP.112–15 SAVILE, D. B. O. 1950. The flight mechanism of swifts and hummingbirds. *Auk* 67: 499–504 (also discusses the alleged beating of wings alternately).

P.112 EDGERTON, H. E., NIEDRACH, R. J. and RIPER, W. V. 1951. Freezing the flight of hummingbirds. *Nat. Geog. Mag.* 100: 245–261.

P.116 WEITNAUER, E. 1947. Am Neste des Mauerseglers, *Apus apus apus* (L.). *Orn. Beob.* 44: 133–182 (flight-speed).

P.116 SLIJPER, H. J. 1948. Over de Gierzwaluw, *Apus a. apus* (L.). *Ardea* 36: 42–51 (flight-speed).

P.116 MEINERTZHAGEN, R. 1921. Some preliminary remarks on the velocity of the migratory flight of birds, with special reference to the palaearctic region. *Ibis* 1921: 228–238 esp. p.232.

P.116 MEINERTZHAGEN, R. 1920. Some preliminary remarks on the altitude of the migratory flight of birds. *Ibis* 1920: 920–936.

P.116 MEINERTZHAGEN, R. 1955. The speed and altitude of bird flight (with notes on other animals). *Ibis* 97: 81-117.

P.116 SPAEPEN, J. & DACHY, P. 1952. Le problème de l'orientation chez les oiseaux migrateurs II. Expériences préliminaires effectuées sur des Martinets noirs, *Apus apus* (L.). *Gerfaut* 42: 54–59 (speed),

P.116 BAKER, E. C. S. 1922. Velocity of flight among birds. *Brit. Birds* 16: 31.

P.118 HEINROTH, M. 1911. Zimmerbeobachtungen an seltener gehaltenen europaischen Vögeln. *Verh. Int. Orn. Kong.* 5: 703–764 (esp. 717–722) (drinking and bathing in captivity).

P.118 TWYMAN, R. S. 1914. Avicultural notes. *Zool.* 4: 18: 152–155 (shivered before taking water from lips).

P.118 SPENNEMANN, A. 1928. *Collocalia esculenta linchi* (Horsf.) *Beitr. Fortpfl. biol.* Vög. 4: 53–58, 98–103 (cave swiftlet bathing through rain).

P.119 VAURIE, C. 1947. Chimney swifts bathing. *Auk* 64: 308–309.

P.119 ALDER, L. P. 1951. Smoke-bathing of swifts. *Brit. Birds* 44: 281.

P.119 *Swifts hitting obstacles*: YARRELL, W. 1876–82. *A History of British Birds* (4th ed. A. Newton) 2: 364–371; Paton, E. R. & Pike, O. G. 1929 *The Birds of Ayrshire* p.90; Joy, N. H. 1930 Fatal collision of swifts, *Brit. Birds* 24: 161.

P.119 HARRISON, J. G. 1949. Some developmental peculiarities in the skulls of birds and bats. *Bull. Brit. Orn. Club* 69: 61–67.

P.120 WORTH, C. B. 1943. Notes on the chimney swift. *Auk* 60: 558–564 (eyelids).

P.120 SAVILE, D. B. O. 1950. The flight mechanism of swifts and hummingbirds. *Auk* 67: 499–504 (alleged alternate wing-beats).

CHAPTER 11

P.123 MARKHAM, J. 1951 in E. M. Nicholson's *Birds and Men* opp. p.224 has excellent photograph of a swift roosting on a vertical wall.

P.123 COUCH, J. 1832. Habits of the swift. *London's Mag. Nat. Hist.* 5: 736–737 (roosted vertically in cage).

P.124 RIVIERE, B. 1897. Roosting of the swift. *Zool.* 4: 1: 511 (under eaves).

P.124 SWAINE, C. M. 1945. Notes on the roosting of certain birds. *Brit. Birds* 38: 330–331 (roosting under window and in occupied bed).

P.124 HATTON, R. H. S. (*in litt.*) (roosting on walls and in room, May 1955.)

P.124 AXELL, H. E. (*in litt.*) (roosting on Dungeness lighthouse).

P.124 NELSON, T. H. SC CLARKE, W. E. 1907. *The Birds of Yorkshire* 1: 262–263 (roosting on windowsills).

P.126 GYNGELL, W. 1897. Common swift roosting in tree. *Zool* 4: 1: 468–469.

P.126 TURNER, E. L. 1930. Swift roosting in a tree. *Brit. Birds* 24:129.

P.126 KOSKIMEES, J. 1950. The life of the swift *Micropus apus* (L.), in relation to the weather. *Ann. Acad. Sci. Fern. A.* IV: 15–16 (roosting in tree).

P.126 COX, R. A. F. 1953. Swift roosting in tree. *Brit. Birds* 46: 414.

PP.125–6 *Roosting in clumps in cold weather:*

TICEHURST, N. F. 1909. *A History of the Birds of Kent* p.225.

CATCHPOOL, T. 1846. Singular habit of the swift. *Zool.* 4:1499–1500.

JESSE, E. 1844. *Scenes and Tales of Country Life* p.169 and 1853 *Scenes and Occupations of Country Life* p.311.

SMITH, F. 1856. Extraordinary effect of sudden cold on swifts. *Zool.* 14: 5249–50.

GURNEY, J. H. 1876. Susceptibility of the swift. *Zool.* 2: 11: 5123.

CORBIN, G. B. 1903. Abundance of swifts (*Cypselus apus*) in South Hants. *Zool.* 4: 7: 266–267.

WATSON, J. B. 1930. Mortality among swifts caused by cold. *Brit. Birds* 24: 107.

KUHK, R. 1948. Wirkung der Regen- und Kälteperiode 1948 auf den Mauersegler, *Micropus apus* (L.). *Vogelwarte* I (15): 28–30 (Konstanz).

BURCKHARDT, D. 1948. Sammelbericht über den Frühlingszug und die Brutperiode 1948. *Orn. Beob.* 45: 222–223 (Basel).

P.127 WEITNAUER, E. 1952. Uebernachtet der Mauersegler, *Apus apus* (L.), in der Luft? *Orn. Beob.* 49: 37–44 (gives key earlier references to night ascents), also 1954. Weiterer Beitrag zur Frage des Nächtigens beim Mauersegler, *Apus apus*. *Orn. Beob.* 51: 66–71, and 1955 Zur Frage des Nächtigens beim Mauersegler, IV Beitrag. *Orn. Beob.* 52: 38–39 (gives references to swifts seen through telescope).

P.128 LONGFIELD, C. SC LOWE, P. B. (*in litt.*) (swifts at night in Cyprus and North Africa.)

P.128 GUÉRIN, G. 1923. La vitesse du vol des oiscaux et l'aviation. *Rev. Franc. d'Ornith.* 15: 74–79 (French airman's account quoted).

P.132 KIRKPATRICK, K. M. 1952. Peculiar roosting site of the house swift (*Micropus affinis*). *J. Bombay Nat. Hist. Soc.* 49: 551–552.

P.132 SALVIN, O. & GODMAN, F. D. 1888–1904. *Biologia Centrali-American*, pp.366–367 (roosting of fork-tailed swift).

P.132 LACK, D. 1956. A review of the genera and nesting habits of swifts. *Auk* 73: 1–32 (brief review of roosting habits of foreign species).

P.133 BENT, A. C. 1940. Life histories of North American cuckoos, goatsuckers, hummingbirds and their allies. *U.S. Nat. Mus. Bull.* 176: 254–316 (roosting of American swifts with detailed accounts of roosting behaviour of chimney swift).

P.132 GROSKIN, H. 1945. Chimney swifts roosting at Ardmore, Pennsylvania. *Auk* 62: 361–370.

P.132 MATHEWS, G. M. 1922–3. *The Birds of Australia* 10: 263 (wood swallows clumping).

P.132 LORENZ, K. 1932. Beobachtungen an Schwalbcn anlässlich der Zugkatastrophe im September 1931. *Vogelzug* 3: 4–10 (swallows clumping).

p.133 KING, I. B. 1941 (roosting of wren) *Oxford Orn. Soc. Rep.* 1940: 12.

CHAPTER 12

PP.135-40 MCATEE, W. L. 1947. Torpidity in birds. *Anter. Midland Nat.* 38: 191–206 (a valuable review of the whole literature; hence only the key references are added below).

P.135 ARISTOTLE (fourth century B.C.). *History of Animals* trs. D'A. W. Thompson 1910.

P.135 WHITE, G. 1795. *The Natural History of Selborne.*

P.135 PENNANT, T. 1768. *British Zoology* (2nd ed.).

P.136 NELSON, T. H. & CLARKE, W. E. 1907. *The Birds of Yorkshire* 1: 262 (torpid at Bolton Hall).

P.136 KLEIN, J. T. 1750. *Historiae avium prodromus* etc. p.204 (the most convincing account outside Britain for hibernating swifts in an old oak, but the date was not specified).

P.136 DREWITT, F. D. 1931. *The Note-book of Edward Jenner* (includes a life of Jenner which in some important points corrects that in the *Dictionary of National Biography*).

P.136 JENNER, E. 1824 (published posthumously). Some observations on the migration of birds. *Phil. Trans. Roy. Soc.* 114: 11–44.

P.137 CAREW, R. 1602. *The Survey of Cornwall* (1st book) p.25 (for quotation from Olaus Magnus.)

P.138 REEVE, H. 1809. *An Essay on the Torpidity of Animals* (for much earlier literature).

P.138 RENNIE, J. 1835. *The Faculties of Birds* (for much earlier literature).

P.138 RYDZEWSKI, W. 1951. A historical review of bird marking. *Dansk Orn. For. Tidsk.* 45: 61–95 (story of early accounts of marked swallows pp.63, 67–68).

P.139 SCOTT, W. 1884. The winter passeres and picariae of Ottawa. *Auk* I: 161 (hibernating chimney swift).

P.139 HANNA, W. C. 1917. Further notes on the white-throated swifts of Slover Mountain. *Condor* 19: 1–8.

P.141 JAEGER, E. C. 1948. Does the poor-will 'hibernate'? *Condor* 50: 45–46. 1949 Further observations on the hibernation of the poor-will. *Condor* 51: 105–109.

CHAPTER 13

P.142 LACK, D. 1955. The summer movements of swifts in England. *Bird Study* 2: 32–40 (gives references to all the previous British observations).

P.145 KOSKIMIES, J. 1950. The life of the swift, *Micropus apus* (L.), in relation to the weather. *Ann. Acad. Sci. Fenn.* A IV Biol., pp.1–151.

P.145 SVARDSON, G. 1951. Swift (Apus apus L.) movements in summer. Proc. X. Int. Orn. Cong., pp.335-338.

P.150 SVÄRDSON, G. 1948–51. JENNING, W. 1953. Verksamheten vid Ottenby fagelstation 1948–52. Fågelvärld 8: 115–116; 11: 172; 12: 162 (ringing recoveries of Swedish birds).

P.151 MATHEW, A. H. 1915. Large flock of alpine swifts in Kent. *Brit. Birds* 9: 95.

P.152 UDVARDY, M. D. F. 1954. Summer movements of black swifts in relation to weather conditions. *Condor* 56: 261–267.

P.152 BRIDGEWATER, A. E. 1934. Notes on the movements of swifts. *Emu* 34: 97–99 (in Australia).

P.152 HOESCH, W. & NIETHAMMER, G. 1940. Die Vogelwelt Deutsch-Sudwest-Afrikas. *J. f. Ornith. 88* (*suppl.*): 202.

P.152 CHAPIN, J. P. 1939. Birds of the Belgian Congo. *Bull. Amer. Mus. Nat. Hist.* 75: 2: 456–458.

CHAPTER 14

P.154 TRISTRAM, H. B. 1867. *The Natural History of the Bible* esp. pp.204–208 (for the Hebrew words).

P.156 WEITNAUER, E. 1947. Am Neste des Mauerseglers, *Apus apus apus* (L.). *Orn. Beob.* 44: 132–188 (departures from Switzerland).

P.157 DARWIN, C. 1871. *The Descent of Man* (quoted from 2nd ed. 1890, p.108).

P.158 SALMON, J. D. 1837. Migration of swifts. *Charlesworth's Mag. Nat. Hist.* 1 n.s.: 108–110 (young in October).

P.161 STRESEMANN, E. 1917. Beobachtungen über die Hohe des Segler-fluges. *Verhb. Orn. Ges. Bayern* 13: 50–52, 278–279 (early height records).

P.161 BERTHET, G. 1934. Note sur le martinet noir *Micropus apus. Alauda* 6: 403–405 (in Lyonnais).

P.161 BOURNE, W. R. P. 1955. Migration seen in the Pyrenees in August and September. *Ibis* 97: 307.

P.161 MITCHELL, K. D. G. 1955. Aircraft observations of birds in flight. *Brit. Birds* 48: 62–63.

P.161 BLEZARD, E. 1943. *The Birds of Lakeland* p.63 (over Carlisle) (20 at 3,400 feet).

P.161 WEITNAUER, E. 1952. Uebernachtet der Mauersegler, *Apus apus* (L.), in der Luft. *Orn Beob*. 49: 37–44; also 1955 Zur Frage des Nächtigens beim Mauersegler, IV Beitrag. *Orn. Beob*. 52: 38–39 (Swiss records up to 6,000 feet up).

P.161 MEINERTZHAGEN, R. 1955. The speed and altitude of bird flight. *Ibis* 97: 107 (further height records).

P.161 PENROSE, H. 1944. Star flight of the swifts. *Country Life* 1944: 593 (one at 7,500 feet). Reprinted 1949 in *I Flew with the Birds* pp.87–91.

P.162 HOLLOM, P. A. D. *in litt.* (migration through southern France, seen with E. M. Nicholson).

P.162 HURRELL, H. G., 1948. Simultaneous watch for migrant swifts, 11th May 1947. *Brit. Birds* 41: 138–145 (migration in progress).

P.162 SLIJPER, H. J. 1948. Over de Gierzwaluw, *Apus a. apus* (L.). *Ardea* 36: 42–51 (departure in the evening).

P.162 British Ornithologists' Club 1906–14. Reports on the immigrations of summer residents in the springs of 1905–13. *Bull. Brit. Orn. Club* vols. 17–34 (reports on migration, arrivals and departures, found dead at lighthouses).

P.162 NELSON, T. H. & CLARKE, W. E. 1907. *The Birds of Yorkshire* 1: 262 (found at lighthouse).

P.162 BARRINGTON, R. M. 1900. *The Migration of Birds as observed at Irish Lighthouses and Lightships* pp.172–175 (dead below lighthouse).

P.162 BAXTER, E. V. & RINTOUL, L. J. 1953. *The Birds of Scotland* 1: 244–246 (killed at lighthouses).

P.163 ALEXANDER, W. B. 1937. Oxford migrant table. *Rep. Oxf. Orn. Soc. for 1936*: 50–54.

P.163 BOYD, A. W. *in litt.*

P.163 OLDHAM, C. 1937. Migratory birds in Hertfordshire. *Trans. Herts. Nat. Hist. Soc.* 20: 141–150.

P.163 TICEHURST, C. B. 1932. *A History of the Birds of Suffolk*, p.193.

P.163 RJVIERE, B. B. 1930. *A History of the Birds of Norfolk,* p.90.

P.163 BERNIS, F. 1951. Sobre el vencejo comun *'Micropus apus apus'* (L.) y su migracion en España. *Bol. Real Soc. Espan. Hist. Nat.* 49:15–40.

P.163 HAARTMAN, L. V. 1951. Die Ankunftszeiten des Mauerseglers, *Apus apus* (L.) und ihre Beziehung zur Temperatur. *Soc. Sci. Fenn. Comm. Biol.* 11: 1–21.

P.163 WHITE, G. 1795. *The Natural History of Selbome.*

P.164 USSHER, R. J. and WARREN, R. 1900. *The Birds of Ireland* pp.102–103 (Miss Elsam, quoted from the Field for 29th May 1886).

PP.164–5 *Published recoveries of ringed birds mentioned in the text are:*

CASTELLARNAU, P. I. S. DB 1949. Catalogo de aves anilladas. *Broteria* (Lisboa) 18: 128–129.

DROST, R. 1933. Der erste Afrikafund eines beringten Mauerseglers (*Apus apus* (L.)). *Vogelzug* 4: 33–34 (from the Vogelwarte Helgoland).

FAKLER, J. 1935. Zweiter Belgisch-Kongo-Fund eines deutschen Mauerseglers (*A. a. apus*). *Vogelzug* 6: 132 (from the Vogelwarte Rossitten).

JÄGERSKIÖLD, L. A. 1937. Goteborgs Naturhistoriska Museums ringmärkningar av flyttfåglar under 1936. *Göoteborgs Musei Arstryck* 1937 p.126.

FONTAINE, V. 1954. Göteborgs Naturhistoriska Museums ringmärkningar av flyttfaglar under 1953. *Göteborgs Musei Arstryck* 1954 p.17.

For the other ringing records specified in the text I am grateful to the Vogelwarte Helgoland (two German recoveries in Spain), the Leiden Museum of Natural History (a Dutch recovery in Austria), the Vogelwarte Radolfzell (a German recovery in French Equatorial Africa) and the Vogelwarte Sempach (a Swiss recovery in the Belgian Congo), and these and other European organisations very kindly furnished me with the full details of their other recoveries of swifts away from where ringed, from which the general summary was compiled.

PP.165-6 LACK, D. 1956. A revision of the species of *Apus. Ibis* 98: 34–62 (summarises knowledge on winter homes of common alpine and pallid swifts, and names of numerous correspondents in South Africa who supplied information on common swifts there).

P.165 ROBERTS, A. 1940. *The Birds of South Africa*, quoted from 2nd ed. 1943 p.154 (wintering in S. Africa).

P.165 HOESCH, W. & NIETHAMMER, G. 1940. Die Vogelwelt Deutsch-Südwest-Afrikas. *J. f. Ornith.* 88 (suppl.): 202 (wintering in S.W. Africa).

P.165 MOREAU, R. E. 1952. The place of Africa in the palaearctic migration system. *J. Anim. Ecol.* 21: 250–271 (migration of wheatear and others).

P.166 LYNES, H. 1925. On the birds of North and Central Darfur. *Ibis* 1925: 360–363 (summering in S. Sudan).

P.166 CHAPIN, J. P. 1939. Birds of the Belgian Congo. *Bull. Amer. Mus. Nat. Hist.* 75: 2: 456–458, 462 (status of common and alpine swifts).

P.166 ARN, H. 1942. Beringungsergebnisse der Alpensegler (*Micropus melba melba* L.). Alter und Nistplatztreue. *Orti. Beob.* 39: 150–162 (recoveries of alpine swifts).

P.166 SCHIHFERLI, A. 1951, 1953. Bericht der Schweiz. Vogelwarte Sempach für die Jahre 1949 und 1950; 1951 und 1952. *Orn. Beob.* 48: 195; 50: 187 (recoveries of Alpine swifts).

P.167 LINCOLN, F. C. 1944. Chimney swift's winter home discovered. *Auk* 61: 604–609.

P.167 ZIMMER, J. T. 1945. A chimney swift from Colombia. *Auk* 62: 145.

P.167–8 BRACKBILL, H. 1950. The man who turned in the first chimney swift bands from Peru. *Migrant* 21:17–21.

CHAPTER 15

LACK, D. 1956. A revision of the species of *Apus*. *Ibis* 98: 34–62 (gives full references to other works.)

PETERS, J. L. 1940. *Check-List of Birds of the World* 4: 244–252 (for previous list).

CHAPTER 16

PP.182–90 LACK, D. & E. 1951. The breeding biology of the swift *Apus apus*. *Ibis* 93: 501–546. (This gives the basic figures for survival in relation to brood-size, but they have been revised, with figures for several further years, in the present chapter.)

P.188 WEITNAUER, E. & LACK, D. 1955. Daten zur Fortpflanzungsbiologie des Mauerseglers (*Apus apus*) in Oltingen und Oxford. *Ornith. Beob.* 52: 137–141 (gives comparison up to 1954 inclusive between Swiss and English birds).

PP.188–92 MOREAU, R. E. 1942. The breeding biology of *Micropus caffer streubelii* Hartlaub, the white-rumped swift. *Ibis* 1942: 27–49. 1941 A contribution to the breeding biology of the palm-swift, *Cypselus parvus. J. East Africa & Uganda Nat. Hist. Soc.* 15: 154–170.

P.189 LACK, D. & ARN, H. 1947. Die Bedeutung der Gelegegrösse beim Alpensegler. *Ornith. Beob.* 44: 188–210 (clutch-size and survival of alpine swifts; H. Am gave *in litt.* additional figures for the critical broods of 4, up to 1953 inclusive, in all 24 young flying out of 44 young in 11 broods, or 55 per cent).

P.192 SERLE, W. 1954. A second contribution to the ornithology of the British Cameroons. *Ibis* 96: 58 (driver ants destroying swifts).

CHAPTER 17

P.193 NICHOLSON, E. M. 1927. *How Birds Live.*

PP.193–94 LACK, D. & ARN, H. 1953. Die mittlere Lebensdauer des Alpenseglers. *Ornith. Beob.* 50: 133–137 (gives Weitnauer's data for common swift as well as Arn's for alpine swift).

P.194 MAGNUSSON, M. & SVÄRDSON, G. 1948. Livslängd hos tornsvalor (*Micropus apus* (L.)). Fågelvärld 7:129–144 (with English summary).

P.194 LACK, D. 1954. *The Natural Regulation of Animal Numbers* (for age of other birds).

P.195–6 UTTENDÖRFER, O. 1939. *Die Ernährung der deutschen Raubvögel und Eulen* (swifts caught by various hawks and owls).

P.196 KLAAS, C. 1953. Zur Ernährung des Baumfalken. *Vogelwelt* 74: 48–49 (swifts caught by hobby mainly in bad weather).

P.196 KUHK, R. 1948. Wirkung der Regen- und Kalteperiode 1948 aut den Mauersegler, *Micropus apus* (L.). *Vogelwarte* 1: 28–30 (caught by birds of prey in cold weather).

P.197 TURNER, E. L. 1920. Kestrel capturing a swift. *Brit. Birds* 14: 136.

P.197 SLIJPER, H. J. 1948. Over de Gierzwaluw, *Apus a. apus* (L.). *Ardea* 36: 42–51 (swifts following hobby; also personal observation on following a sparrowhawk, and for following a buzzard, letter from J. H. Owen filed at Edward Grey Institute, Oxford).

P.197 MORSHEAD, P. E. A. 1929. Hobby and swift. *Brit. Birds* 23; 64,

P.197 WHITE, GILBERT 1795. *The Natural History of Selborne.*

P.197 WEITNAUER, E. 1947. Am Neste des Mauerseglers *Apus apus apus* (L.). *Orn. Beob.* 44: 133–182 (weasel destroying young in nest, adults alarmed when he caught young, adults with *Crataerina*).

P.197 MACPHERSON, H. A. 1897. *A History of Fowling* p.155 (swift towers in Tuscany).

P.197–8 WATERTON, C. 1844. *Essays on Natural History, 2nd ser.* vol. 2 quoted from new ed. 1857) p.lxxiv (swifts in Rome).

P.198 MATHEW, M. A. 1867. Capture of swifts by hook and line. *Zool.* 2: 2: 827.

P.199 KEMPER, H. 1951. Beobachtungen an *Crataerina pallida* Latr. und *Melophaga ovinus* L. (Diptera Pupipara). *Zeit. Hyg. Zool. Schädl. Bekämpf.* 39: 225–259.

P.199 BÜTTEKER, W. 1944. Die Parasiten und Nestgäste des Mauerseglers (*Micropus apus* L.). *Orn. Beob.* 41: 25–35.

P.200 GIMINGHAM, C. T. 1938. A swift infested by parasites. *Trans. Herts. Nat. Hist. Soc.* 20: 352.

P.200 ROTHSCHILD, M. & CLAY, T. 1952. *Fleas, Flukes and Cuckoos* p.150 (feather-lice of swift).

P.201 BEEBE, W. 1949. The swifts of Rancho Grande, North-central Venezuela, with special reference to migration. *Zoologica* 34: 53–62 (black swifts taken by falcon).

P.201 SUTTON, G. M. 1936. *Birds in the Wilderness* p.182 (merlin took Vaux swift).

P.201 CHAPIN, J. P. 1939. Birds of the Belgian Congo. *Bull. Amer. Mus. Nat. Hist.* 75: 2: 447 (peregrine taking *Chaetura cassini*).

P.201 LANE, F. W. 1948. *Animal Wonderland* p.50 (bass taking chimney swift).

CHAPTER 18

P.205 BUTLER, SAMUEL 1879. *Evolution, Old and New.*

P.205 SHAW, BERNARD 1928. Preface to *Back to Methuselah.*

P.206 FISHER, R. A. 1930. *The Genetical Theory of Natural Selection*; 1950. *Creative Aspects of Natural Law* 4th A.S. Eddington Memorial Lecture (Cambridge); 1954 Retrospect of the criticisms of the theory of natural selection, in *Evolution as a Process* (ed. J. S. Fluxlcy) pp.84–98.

P.207 PUSEY, E. B. 1878. *Un-science not Science, adverse to Faith*: a sermon preached before the University of Oxford. Notes p.52.

P.207 LACK, D. 1947. *Darwin's Finches.*

P.207 WADDINGTON, C. H. 1942. *Science and Ethics.*

P.207 SIMPSON, G. G. 1950. *The Meaning of Evolution* (evolution of ethics).

P.207 THORPE, W. H. n.d. (1951.) Evolution and Christian Belief. *Brit. Soc. Biol. Council Occ. Pap. 7.*

P.207 DARWIN, F. (ed.) 1887. *The Life and Letters of Charles Darwin* 1: 316.

CHAPTER 19

THE TOWER, P.210–213

Bromhall, D. (1980) *Devil Birds: the Life of the Swift*. Hutchinson. London.

Lack, A. & Overall, R. (2002) *The Museum Swifts*. Oxford University Museum of Natural History.

Overall, R. (2015) Guardian of the swifts in the tower of the Oxford University Museum of Natural History. *Fritillary (Journal of the Ashmolean Natural History Society of Oxfordshire and the B.B.O.W.T)*, 6, 77–93.

CLASSIFICATION, P.214

Brooke, R.K. (1968) *Apus berliozi* Ripley, its races and siblings. *Bulletin of the British Ornithologists' Club*, 89, 11–16.

Chantler, P. & Driessens, G. (2000) *Swifts: A Guide to the Swifts and Treeswifts of the World*, 2nd ed. Pica Press, Sussex.

de Juana, E. & Garcia, E. (2015) *The Birds of the Iberian Peninsula*. Bloomsbury, London.

Ksepka, D. T., Clarke, J. A., Nesbitt, S. J., Kulp, F. B. & Grande, L. (2013) Fossil evidence of wing shape in a stem relative of swifts and hummingbirds (Aves, Pan-Apodiformes). *Proceedings of the Royal Society B*, 280, DOI: 10.1098/rspb.2013.0580.

Leader, P. J . (2011) Taxonomy of the Pacific Swift, *Apus pacificus* Latham, 1802, complex. *Bulletin of the British Ornithologists' Club*, 131, 81–83.

Mayr, G. (2002) Osteological evidence for paraphyly of the avian order Caprimulgiformes (nightjars and allies). *Journal fur Ornithologie*, 143, 82–97.

Ruf, T. & Geiser, (2015) Daily torpor and hibernation in birds and mammals. *Biological Reviews of the Cambridge Philosophical Society*, 90, 891-926. doi: 10.1111/brv.12137

WorldBirdNames.org (2012) *Swifts, Hummingbirds and allies*. version 2.11 http://www.worldbirdnames.org/n-swifts.html; downloaded August 2017.

NESTING, P.220

Martins, T. L. F. (1997) Fledging in the common swift, *Apus apus*: weight-watching with a difference. *Animal Behaviour*, 54, 99–108.

Martins, T. L. F. & Wright, J. (1993a) Cost of reproduction and allocation of food between parent and young in the swift (*Apus apus*). *Behavioral Ecology*, 4, 213–223.

Martins, T. L. F. & Wright, J. (1993b) Brood reduction in response to manipulated brood sizes in the common swift (*Apus apus*). *Behavioral Ecology and Sociobiology*, 32, 61–70.

Midway Atoll National Wildlife Refuge (2016) Wisdom, the Laysan Albatross. https://www.fws.gov/refuge/Midway_Atoll/wildlife_and_habitat/Wisdom_Profile.html downloaded, August 2017.

Overall, R (2015) – see under *The Tower*.

Perrins, C. M. (1971) Age of first breeding and adult survival in the swift. *Bird Study*, 18, 61–70.

Thomson, D. L., Douglas-Home, H., Furness, R. W. & Monaghan, P. (1996) Breeding success and survival in the common swift *Apus apus*: a long term study on the effects of weather. *Journal of Zoology, London*, 239, 29–38.

SCREAMING AND BANGING, P.223

Ansorge, K. (2015) Sexual dimorphism of acoustic signals in the Common Swift *Apus apus*. *APUSlife* 2015, no. 5457 www.commonswift.org/5457Ansorge.html.

Olos, G. (2017) Is 'banging' an antipredator behaviour in Common Swift (*Apus apus*)? *Ornis Fennica*, 94, 45–52.

PARASITES, P.225

Bize, P., Roulin, A., Tella, J., Bersier, L.-F. & Richner, H. (2004) Additive effects of ectoparasites over reproductive attempts in the long-lived Alpine swift. *Journal of Animal Ecology*, 73, 1080–1088.

Walker, M.D. & Rotherham, I.D. (2010) The breeding success of Common Swifts *Apus apus* is not correlated with the abundance of their louse-fly *Crataerina pallida* parasites. *Bird Study*, 57 (4) http://dx.doi.org/10.1080/00063657.2010.493581

Walker, M.D. & Rotherham, I.D. (2011) No evidence of increased parental investment by Common Swifts *Apus apus* in response to parasite loads in nests. *Bird Study*, 58 (2) http://dx.doi.org/10.1080/00063657.2010.546391

FLIGHT, P.226

Del Hoyo, J., Elliott, R. & Sargatal, J. (eds.) (1999) *Handbook of the Birds of the World, vol 5 Barn-owls to Hummingbirds*. Lynx Edicions, Barcelona.

Hedenström, A. (1998) The stoop of large falcons. *Trends in Ecology and Evolution*, 13, 383–385

Hedenström, A. & Åkesson, S. (2017) Adaptive airspeed adjustment and compensation for wind drift in the common swift: differences between day and night. *Journal of Animal Behaviour* DOI:10.1016/j.anbehav.2017.03.010

Henningsson, P. & Hedenström, A. (2011) Aerodynamics of gliding flight in common swifts. *Journal of Experimental Biology*, 214, 382–393.

Henningsson, P., Johansson, L.C. & Hedenström, A. (2010) How swift are swifts *Apus apus*? *Journal of Avian Biology*, 41, 94–98.

SLEEP, P.228

Bäckman, J. & Alerstam, T. (2002) Harmonic oscillatory orientation relative to the wind in nocturnal roosting flights of the swift *Apus apus*. *Journal of Experimental Biology*, 205, 905–910.

Hedenström, A., Norevik, S.G., Warfvinge, K., Andersson, A., Bäckman, J. & Åkesson, S. (2016) Annual 10-month aerial life phase in the common swift *Apus apus*. *Current Biology*, 26, 3066-3070. http://dx.doi.org/10.1016/j.cub.2016.09.014

Liechti, F., Witrilet, W., Weber, R. & Bachler, E. (2013) First evidence of a 200-day non-stop flight in a bird. *Nature Communications*, 4, 2554. doi:10.1038/ncomms3554

Rattenborg, N.C. (2017) Sleeping on the wing. *Interface Focus*, 7, 20160082 http://dx.doi.org/10.1098/rsfs.2016.0082

Tarburton, M.K. & Kaiser, E. (2001) Do fledgling and pre-breeding common swifts, *Apus apus*, take part in aerial roosting? An answer from radio tracking equipment. *Ibis*, 143, 255-263. Doi:10.1111/j.1474-919X.2001.tb04481.x

MIGRATION, P.230

Åkesson, S., Bianco, G. & Hedenström, A. (2016) Negotiating an ecological barrier: crossing the Sahara in relation to winds by common swifts. *Philosophical*

Transactions of the Royal Society B Biological Sciences, 371, issue 1704. DOI: 10.1098rstb.2015.0393

Åkesson, S., Klaassen, R., Holmgren, J., Fox, J. W. & Hedenström, A. (2012) Migration routes and strategies in a highly aerial migrant, the common swift, *Apus apus*, revealed by light-level geolocators. *PLoS One*, 7(7): e41195. doi:10.1371/journal.pone.0041195.

Appleton, G. (2012) Swifts start to share their secrets. *British Trust for Ornithology News*, May-June 2012, 16-17. https://www.bto.org/sites/default/files/u49/BTO_299_16-17Swifts

Townshend, T. (2015) Out of Africa! The Beijing swift's incredible journey charted at last. https://birdingbeijing.com/2015/05/24/. Accessed12 September 2017.

NUMBERS IN THE TOWER, P.233

Overall, R. (2015) see under *The Tower*.

Personal communication from Roy Overall, George Candelin, Andrew Gosler and Sandra Bouwhuis.

SWIFT CONSERVATION, P.237

Action for Swifts (2013) How many swifts are there in GB and UK? http://actionforswifts.blogspot.co.uk/2013/02/how-many-swifts-are-there-in-gb-and-uk.html.

Balmer, D. E., Gillings, S., Caffrey, B. J., Swann, R. L., Downie, I. S. & Fuller, R. J. (2013) *Bird Atlas 2007–11: the Breeding and Wintering Birds of Britain and Ireland*. BTO Books, Thetford.

Benton, T. G., Bryant, D. M., Cole, L. & Crick, H. Q. P. (2002) Linking agricultural practice to insect and bird populations: a historical study over three decades. *Journal of Applied Ecology*, 39, 673–687.

Evans, K. L. and Bradbury, R. B. (2007) Effects of crop type and aerial invertebrate abundance on foraging barn swallows, *Hirundo rustica*. *Agriculture, Ecosystems and Environment*, 122, 267–273.

Evans, K. L., Newton, J., Mallord, J. W. & Markman, S. (2012) Stable isotope analysis provides new information on winter habitat use of declining avian migrants that is relevant to their conservation. *PLoS One*, 7(4):e34542 https://doi.org/10.1371/journal.pone.0034542.

Fitzgerald, T. M., van Stam, E., Nocera, J. J. & Badzinski, D. S. (2014) Loss of nesting sites is not a primary factor limiting northern Chimney Swift populations. *Population Ecology, Tokyo*, 56, 507–512.

Hayhow, D. B., Ausden, M.A., Bradbury, R. B. & 13 others (2017) The state of the UK's birds, 2017. *RSPB, BTO & 5 others*, Sandy, Bedfordshire.

Nebel, S., Mills, A., McCracken, J. D. & Taylor, P. D. (2010) Declines of aerial insectivores in North America follow a geographic gradient. *Avian Conservation and Ecology*, 5:1 http://dx.doi.org/10.5751/ACE-00391-050201

INDEX

Italic numbers indicate illustrations. **Bold** numbers indicate tables.